DICTIONARY OF
ENTOMOLOGY

First Published, 2010
Reprinted, 2026

ISBN: 978-81-7035-650-9

Publisher's Note:

Every possible effort has been made to ensure that the information contained in this book is accurate at the time of going to press, and the publisher and author cannot accept responsibility for any errors or omissions, however caused. No responsibility for loss or damage occasioned to any person acting, or refraining from action, as a result of the material in this publication can be accepted by the editor, the publisher or the author. The Publisher is not associated with any product or vendor mentioned in the book. The contents of this work are intended to further general scientific research, understanding and discussion only. Readers should consult with a specialist where appropriate.

Every effort has been made to trace the owners of copyright material used in this book, if any. The author and the publisher will be grateful for any omission brought to their notice for acknowledgement in the future editions of the book.

Published by : **Daya Publishing House®**
A Division of
Astral International Pvt. Ltd.
– ISO 9001:2015 Certified Company –
4736/23, Ansari Road, Darya Ganj
New Delhi-110 002
Ph. 011-43549197, 8130496929
E-mail: info@astralint.com
Website: www.astralint.com

Printed at : **Replika Press Pvt. Ltd.**

Preface

Words in any language have always taken a major role in the advancement of human civilization. With the passing of time many words automatically added and enriched the various cultures towards provisions of a better, secured and longer life of all the living beings on this planet–the earth. Words have been used in the area of education which in turn has played a significant role towards improvement and upgradation of human life. In this regard scientific knowledge acquired through modern scientific research and education has influenced the human life tremendously vis-à-vis all other connected human activities.

Agricultural science including animal science as a particular way of life as adopted by human beings and also moved in the same direction as other sciences and also acts made good progress only primarily by means up many words with passing out of time. As far as scientific development for further improvement is concerned, it firstly

known as pleuron. The hinder segments are hidden so that the eighth and ninth terga of the female and the ninth tergum of the male are largely hidden by the seventh tergum.

Abdomen of culex It is long and slender, and the abdominal segments bear paired tuft of bristles. The eighth segment carries a long conical respiratory siphon, which the larva must thrust into air at intervals to breathe.

Abdomen of housefly It is oval and is composed of eight segments in the male and nine in the female. The first segment is usually atrophied and the second is very much reduced. In the female Musca, the visible segment forms the retractile ovipositor. The abdomen is also covered with fine hair or bristles.

Abdomen of insects It bears eleven or fewer segments and also the genital apertures, which are situated near the anus at the posterior end of the body.

Abdomen of mosquito It is long and narrow except at the end of a good meal when it is swollen with blood. The individual segments are then visible, somewhat easily. There are not true appendages. There are nine segments altogether and these bear, however, no knobs at the end of the abdomen which are related to egg laying in the female. The anus lies at the end of the eighth segment.

Abductor and adductor muscles The mandibles of an insect are moved in the horizontal plane by the alternate contraction of the abductor and adductor muscles.

Acanthiophilus helianthi Rossi *See* safflower fly.

Acantho Thorn.

Achoea janata Linn *See* fruit sucking moth of citrus crops and Castor semi-looper of Castor.

Adoretus duvauceli **Bl** *See* leaf and fruit eating beetles of apple.

Adoratus lasiopygus Commonly known as Cockchafer beetle, defoliates the grapevine at night. Another species is A. *versutus*. May be controlled by means of light traps or stomach sprays of insecticide.

Adult locust Adult locust flying in swarms, are the most difficult to control. If they alight on any field or ground, they should be immediately killed in as many numbers as possible. This is best done in the mornings when they are lazy. If the swarm is of pink coloured locusts, they should be dusted with 10 per cent B.H.C. Aerial spray of flying swarms and ground hoppers have to be restored too with aldrin or dieldrin solutions in oil in case of very serious infestation.

Aedes aegyptica The common yellow fever mosquito. The cause of the diseae is an extremely minute organism or filterable virus which passes an essential part of its life cycle in each host–man and mosquito.

Agromyza obtusa **M.** *See* gram pod fly of gram.

Agromyza phaseoli **C.** *See* stem-fly of cowpea.

Agromyzid flies Attack pea, arhar, mung, urid, lentil, etc.; maggots bore into pods and seeds.

Agrotis **spp.** *See* cutworm of linseed.

Agrotis ypsilon Rott. *See* greasy cutworm of gram.

Air tubes in culex The siphon contains two air tubes which are connected with a system of trachea inside the body of larva. The supply of air through trachea is supplemented by forefinger like tracheal-gills.

Alary muscles The alary muscle of the cockroach is triangular with its broad end near the middle of the pericardium.

Alcides mali Marshal *See* shoot borer of apple.

Aleurocanthus 1-uesaini* and *A. woglumi A black spiny fly, is a serious insect of citrus crop. All these flies can be controlled by a spray of rosin soap in the winter and rosin compound in September. Tobacco decoction has also been reported to be effective to control this insect.

***Aleurocanthus spiniferus* Quaint** See white flies of citrus crops.

***Aleurolobus barodensis* M** *See* Mealy wings of sugarcane.

Aleurolobus citrifolia The white fly of citrus is serious pest.

Alimentary canal of cockroach It is a slightly coiled canal and consists of three parts (*i*) a long stomodeum, (*ii*) a short mesenteron, (*iii*) a long, slightly coiled proctodeum.

Almond moth, *Ephestia cautella* Walker A grey moth whose pinkish white caterpillar feeds on all sorts of dry fruits, chocolate, cocoa bean, biscuits, cereal grains, flours, oil seeds, garlic, tamarind seed, etc.; eggs are laid on food materials or anywhere in their vicinity, caterpillar lives in small tubes made by it by gluing together particles of food materials with a silky material secreted by it.

Ambulacral Pertaining to the tube feet of an echinoderm.

Amsacta moorei **Butler** *See* Red hairy caterpillar of sann-hemp.

Amoeboid cells In insects, in addition to the excretory role of fat body, it has been seen that a good bit of waste matter is also deposited beneath the cuticle by the wandering amoeboid cells and is ultimately incorporated in the chitin.

Anal cerci of cockroach In both the sexes, the tenth tergum projects backwards as a plate with a deep notch in the hinder edge. A pair of spindle shaped, many jointed anal cerci, attached under this plate and below is the anus between two podical plates or paraprocts, which are believed to represent the sternum of the eleventh somite.

Anal styles In cockroach, these are the pairs of slender, unjointed structures situated below the ninth sternum and are wanting in the female.

Anaplasmosis This is an infectious disease whose aetiology and symptomology are similar to cattle tick-fever, except that more carriers are involved. It is caused by minute parasite *Anaplasma marginale* which invades the red blood cells. The parasite is transmitted from infected to healthy animals. In infected animals, the protozoa live in the red blood cells. The symptoms may be mild, acute or chronic conditions. The more characteristic symptoms in mature animals include rapid, pounding heart action, laboured and difficult breathing dry muzzle, marked depression, tremors of muscles, loss of appetite and a great reduction in the milk flow. The most effective measures consist of (1) Using the vaccine. (2) Feeding an antibiotic occasionally at low level to control the spread. (3) Spraying approved insecticides in livestock buildings and on cattle to control, kill and ward off biting insects that spread the disease. (4) Preventing its spread thoroughly before using on the next animal.

Anar butterfly of pomegranate (*Punica granatum* L.) *Virachola isorates* Fabr. The butterfly lays eggs on flowers and young fruits, and the caterpillars bore into fruits and feed on seeds and tissues; they pupate in fruit. Pack young fruits with stiff paper or polythene (300 gauges) or coarse cloth; spray repeatedly with 0.25 per cent DDT @ 1–2 gallons per tree at flowering time.

Animal pests Animals and birds are serious pests of fields and orchards. They can be prevented by various means such as by building masonry walls, barbed wire fences or thorny strong hedges and also physically watching the crop fields. In case of monkeys, to rare dogs can be helpful to a great extent.

***Anomala polita* Bl** *See* leaf and fruit eating beetles of apple.

***Anomis sabulifera* Guen** *See* jute semilooper of June.

Anoplura The order of insects that contains the sucking lice *e.g. Pediculus humanus* (the head body and clothes lice). Lice, ectoparasites of mammals, are found universally in unhygienic conditions and are carriers of typhus and other diseases. They lack wings and eyes and have a flattened transparent body with piercing and sucking mouth parts for feeding on the host's blood and pretensile clawed legs for attachment to the host's hair and develop into blood-sucking nymphs.

Antenna One of a pair of threadlike appendages on the head of many arthropods. They generally have a sensory function (touch and smell) but some crustaceans use them for swimming and attachment. *See* also antennule.

In insects, these are long slender, filiform, and many jointed structures, are attached slightly anterior and ventral to the eyes. The base of each antennae is movably fixed into a small antennal socket. Each antennae consists of a broad basal segment known as the pedicel and a long jointed flagellum. Being sensory in function, they sweep the ground when the insect is moving in search of food and shelter. They are also used for search of food, when the insect sweeps the ground with their long antennae which have organ of smell.

Antennal socket The base of each antennae is fixed into a small antennal socket. Each antenna consists of a broad basal segment, the scape, a short cylindrical second segment known as the pedicel, and a many jointed flagellum. Being sensory in function, they sweep the ground when the cockroach is moving in search of food and shelter.

Antennule One of a pair of small threadlike sensory appendages that occur anterior to the antennae on the heads of Crustacea.

Antheraraea myllitta This is the name of a most famous breed of silk moth, also called Tussore. It spins a remarkable cocoon with a long stiff stalk which ends in a loop tightly fastened round a twig, and a number of cocoons hanging in a tree gives an appearances of fruits.

Anticoagulin This is a ferment produced during mosquito bite and is supposed to prevent the blood from clotting and blocking the proboscis.

Antigastra cutalaunalis The leaf and capsule borer of sesame. The population of the pest is governed by four braconid natural enemies *viz. Bracon brenicornis, B. hebter, B. gelechi agathis* spp. and *Sichnumnidsi i.e. Erioborus* spp., *Trathola biguthula, Complexlas* spp.

Ants (1) *Camponotus compressus* Fabr. (2) *Monomorium indicum* Forell (3) *Dorylus* spp. (4) *Holcomyrmex scabriceps* Mayr. Ants infest stores and godowns to feed on grains, flour and other edible products; they also carry them to their nests; most of them are non-injurious to crops.

Anuraphis helichrysi **Kalt** *See* leaf curling anhid of peach.

Aonidiella aurantii The true Californian scale insect.

Aonidiella oeientalis The oriental yellow scale insect of citrus.

Aorta *See* circulatory system in insects. Also *See* heart of cockroach.

Aphids The most general method of transmission of plant diseases is by means of sucking insects, of which aphids, jassids and white flies are the most important. In the field, the dissemination is chiefly through their agency, and it is to a large extent true that if there were no sucking insects to serve as vectors, there would be no virus diseases. These are Hemipetrous insects.

Aphid of citrus *(Texopeera aurantii)*. Can be controlled by fish oil rosin soap.

Aphid of grape vines *(Phylloxera vastatrix)*. Use Bordeaux mixture to control the insect. Calcium arsenate can also be successfully used for control of this insect.

Aphid of linseed (*Linum usitatissimum* L.) *Myzus persicae* Sulz. Aphids congregate in small colonies; plant growth stunted. Spray with 0.2 per cent BHC or 0.05 per cent nicotine sulphate/malathion/fish oil rosin soap @ 40–50 gallons per acre.

Aphid of maize (*Zea mays* L.) *Rhopalosiphum* (Aphis) *maidis* (Fitch). A black aphid on leaves and ears; damage similar to wheat aphid. Same as for wheat aphid.

Aphid of potato (*Solanum tuberosum* L.) *Myzus persicae* Sulz. Attacks potato sprouts and is a vector of several virus diseases. Spray with 0.06 per cent nicotine sulphate or 0.025 per cent parathion @ 60–80 gallons per acre.

Aphid of wheat crop These sucking insects start damaging the wheat and barley crops at an early stage. These include *Rhophalosephum maidis, R. padi, Sitobion avenae, Miscanthii, Soezophis gramineuum, Longinus saccharis, Acyo thesiphum pisum, Myzusp ersaceae* and *Hysteroneuris octoniceae* attach wheat and barley crops. Brown wheat mites are also effective in controlling these.

***Aphis craccivora* K.** *See* cotton aphid of cowpea.

Apholinus citrina A yellow scale insect of citrus.

Apholinus mali An insect parasite of apple aphids that successfully controls the pest.

Apocrita Another division of Hymenoptera includes all the higher forms, which are characterised by the wasp-waist which is lacking in symphyta. Formicoideae (ants) is a sub-group of the Apocrita and all its members are social forms. The sub-groups Vespoidea (wasps) includes three wasps which may be solitary or social forms. The sub-group Apoidea includes honeybees, which may include both solitary and social forms.

Apoidae *See* apocrita.

Apple San lose scale *See* San lose scale of apple.

Apple wooly aphis *See* Wooly aphis.

Apposition image The image which is contributed by a single ommatidium.

***Apsylla cistellata* Buckton** *See* mango shoot gall maker of mango.

Arachnida The class of the Arthropoda that contains the mostly terrestrial and carnivorous scorpions and spiders, which typically spinnerets on the abdomen of web spinning, and the parasitic ticks and mites.

Arboreal pertaining to life on trees.

Archenteron A gastrular cavity in a metazoan embryo which serves as a primitive gut.

Argasidae The common fowl tick. *See* Animal Science.

Argus persicus A tick insect of poultry.

Arolium *See* legs of cockroach.

Armyworms *Mythimna (Pseudoletia) separata* is injurious to wheat crop. Controlled by foliar spray of famiotiothin or dichlorivos @ 555 ml. a.i./Ha or Carbaryl-kg/Ha.

Arthopodal body Segmented and jointed externally.

Arthropods Insects are such arthropods in which body is divided into three regions head–thorax and abdomen.

Arthropoda The phylum (arthros-joint; podos-foot) is the largest group of the animal kingdom, surpassing in the number of species of all the other phyla combined. It contains over 700,000 species. The animals belonging to this phylum occur all over the globe at an altitude of over 6,500 metres on mountains to a depth of over 6,000 metres in the sea. Different species are adapted for life in the air, on land, in soil and in fresh or salt waters.

Asperate lobe *See* titilator.

Aspidiosu destructor A scale insect pest of mango, is a serious pest as it sometimes attacks the tender parts of the tree. Also attacks banana and guava crops.

Aspidiotus cydoniae **Comst.** Nymphs and adult females suck the sap. Spray with fish oil rosin soap (1 part in 40 parts of water) or 0.03 per cent parathion @ 150–200 gallons per acre.

Aspidomorpha miliaris **F** *See* leaf beetles of sweet potato.

Aspondylia sesami **F** *See* gall fly of sesame.

Aspongopus janus **F** *See* bugs of cucurbits.

Aspongopus brunmeus *See* bugs of cucurbits.

Athalia proxima **K** *See* mustard saw-fly of crucifers.

Attacus atlas It is one of the giant silk moths and is the largest of all the living insects reaching upto nearly 27 cms in wing span.

B

Bactrocera persicae **B** *See* fruit fly of custard apple.

Bagrada cruciferarum **Kirk** *See* painted bug of crucifers.

Bagworm of tea (*Camellia sinensis* Kuntze) *Clania crameri* West. Caterpillars defoliate the plants. Spray with 0.16 per cent DDT in 60–8 gallons of water per acre. They also destroy banana crops.

Bark eating caterpillar of citrus crops *indarbela quadrinotat Stalk.* Control Same as for mango. *Inderbela tetoronus.*

Basement membrane The body of cockroach has three distinct layers: (1) Cuticle, (2) Epidermis, and (3) Basement membrane.

Batocera ruformaculata A long horned beetle stem borer of mango, figs and jack fruit. The grubs should be scooped with a knife; destroy with a stiff wire later then. The grubs can be killed by putting carbon disulphide, tar etc. in the hole.

Bees Many types of bees bring about pollination in plants and produce honey; thus economically important.

Beetle borer of citrus *Inderbela quadrininotata.*

Beetle borer of oranges *Stromabum barbatim.* To control this pest, inject little petrol or a mixture of two parts of chloroform to one part of creosote in the hole and close it. Painting the galleries with Paris Green can also be used to control this pest.

Bihar hairy caterpillar of sesame *Spilosoma obliqua.* It is also controlled by the natural parasites.

Bilaterally symmetrical A system in which the body can be equally divided in one plan into two identical halves, as in arthropods.

Biological regulation *See* Crop Science.

Birds Damage crops right from the time of sowing to harvesting and processing. Pigeons (*Collumula livia*), doves (*Streptopalea* spp.), crows (*Corcus* sp.), sparrows (*Passer corcus* sp.) and parakeets all damage our crops in one form or other.

Biting housefly Also called stable-fly or dog-fly. This fly has been incriminated in the mechanical transmission of anthrax, swampy fever, surra and even infantile paralysis in man. It is also known to be in intermediate hoist for the peritoneal roundworm of cattle and the small mouth stomach worm of horses. The destruction of the natural breeding grounds of the biting fly is the first and primary means of prevention. Sprays are the second best means of controlling this pest.

Blowfly All the important species of blowflies except the flesh-flies, which are greyish and have three dark stripes on their backs have more less metallic lustre. It causes the greatest damage by infesting wounds and soiled hair or fleece of living animals. The black blowfly (or wool-maggot fly) is similar to that caused by screw-worm. Sheep are especially susceptible to attacks of blowflies because their wool frequently becomes soiled or moistened by rain and accumulation of faeces and urine. Its prevention of damage consists of eliminating the pest and decreasing the susceptibility of animals to infestation. The practing of docking lambs and tagging sheep at intervals will materially lessen blowfly damage in this class of animals. When animals become infested with blowfly maggots, their wounds should be treated twice weekly with (1) Formula MS 62 or (2) EQ-335 Screw-worm remedy diluted one part to nine parts of water-same treatment as given in screw-worm maggots. Also chloroform is frequently used to kill the larvae.

Body of insects Insects are arthropods in which the body is divided into three regions, the head, thorax and abdomen.

Bristles *See* legs of honeybee.

Brood comb *See* honeycomb.

Brown wheat mite *Tetrbia latens.* This mite can be controlled by a foliar spray of formithion, phosphamidon, oxydiemiton or monocrotophos @ 160, 250, 160 and 200 ml/gallons of water respectively.

Bruchus analis **Fabr.** *See* pulse beetles.

Brush border *See* malpighian tubule.

Bud worm of brinjal (*Solanum melongena*) *Gnorinosochinui* (*Ghthhorineae*) *glapaigona*. A small caterpillar pale white or pinkish, bres in and feeds in flower buds, the moth is small and dirty brown. Control same as for shoot and fruit borer of brinjal.

Bugs of cucurbits *Aspongopus janus* F. and *Brunmeus* N. Red and pale brown insects that suck the sap. Destroy by hand-picking or by catching in nets. Dust with 2–5 per cent BHC @ 15–20 lb. per acre.

Butter fly The characteristics of this insect are (*i*) Antennae are clubbed, (*ii*) The body is slender and not very hairy, (*iii*) In resting position of wings, they are vertical and closed or horizontally expanded, (*iv*) Fly by day, (*v*) Eggs usually barrel shaped, and (*vi*) Pupa angular and often brightly coloured.

***Buzura suppressaria* G** *See* tea looper of tea.

C

Calliphoridae and Sarcaphadae Blow flies and flesh flies. *See* Animal Science.

Caterpillar A stage in metamorphosis of butterflies and moths. The cylindrical body of the caterpillar consists of a head and about thirteen segments. On the head, there are powerful jaws which move transversely and help them in feeding on plant leaves. Their perception of surroundings is by a number of simple eyes (ocelli) on their heads and by the two sensitive hair-like antennae. Inside the head, a pair of salivary glands form the silk producing organs and their ducts project in a tube, the spinnerets behind the jaws. *See* metamorphosis.

Cattle lice The louse is a small flattened, wingless insect parasite of which there are several species, may range from one-fourth inch in length. The two main types are sucking lice and biting lice. Of the two groups,

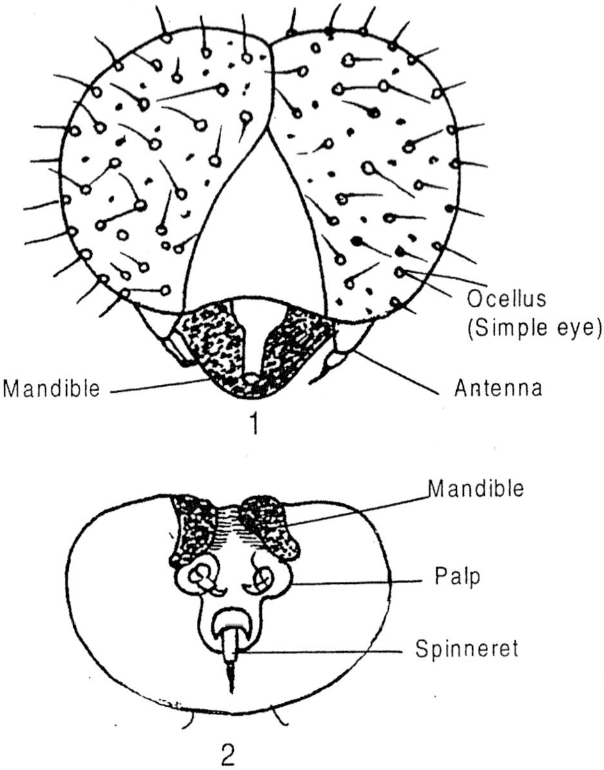

Caterpillars Head
1: Front view; 2: From below

sucking lice are the most injurious. Most species of lice are specific for a particular class of animals. Thus cattle harbour a number of common species of lice including the long nosed and short nosed sucking type of lice and the little red biting lice, but cattle lice will not remain on the other farm animals, nor will lice from other animals infest cattle. Lice are more abundant on weak, unthrifty animals and more troublesome during the winter months than during the rest of the year.

Cattle tick fever Also known as Splenetic fever. This is an infectious protozoan disease of adult cattle caused by a protozoa called *Babesia bigemina* which helps upon the tick chiefly, *Boophilus annulatus* for the survival and transmission. The most effective control measures are directed at the eradication of the fever ticks, either by killing them the pastures on the cattle. Pastures may be rendered free of ticks by excluding all the host animals–cattle, horses and mules–until all the ticks have died of starvation in eight to ten months. Spray with Pyrethrum or Rotonone the affected animals body.

Cephalic The larva of housefly consists of anterior cephalic or head segments and 12 trunk segments.

Cephalothorax The body of the pupa of Culex mosquito consists of a large cephalothorax, that is head plus thorax; and a slender abdomen which is more or less bent under the cephalothora.

Caratopoganoidae *See* Animal Science. Biting midges.

Cereal thrips in maize *Anophotthrips indanesis.* For control, carboferon 36 @ 150g a.e./hec. and 75g *a.e.* seed dressing in sandy loam clay soil.

Cerci *See* anal cerci.

Cervix of insect *See* head of insect.

Chafer beetle of grape-vine (*Vitis Vinifera* L.) *Adoretus lasiophygus* B. The beetle comes at night and feeds on leaves. Spray with 0.25 per cent DDT @ 150–200 gallons per acre.

Chicken mite attacks chicken only.

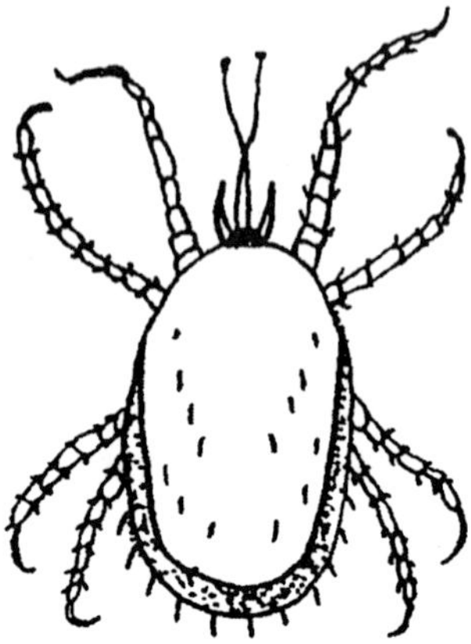

Chicken Mite (Deamanyssus)

Chilli thrips of chilli (*Capsicum futescens* L.) *Scirtothrips dorsalis H.* Nature of damage is the same as in the case of onion and garlic thrips. Spray with 0.1 per cent BHC @ 40–60 gallons per acre. Dust with 5 per cent BHC @ 15–20 lb. per acre.

Chilopoda The class of arthropods that contains the centipedes (*e.g*. Lithobius), characterised by a flat body divided into numerous segments, each bearing one pair of walking legs. They are terrestrial and breathe air through tracheae, excretion by Malpighian tubules. Centipedes are carnivorous, with poison claws on the first body segment. They are sometimes placed with the millipedes (diplopoda) in the group Myriapoda. Compare Diplopoda. *See* Myriapoda.

***Chilotrae infuscatellus* Sn** *See* stem borers of sugarcane.

***Chilo tumidicostalis* Hmps** *See* stem borers of sugarcane.

***Chilo zonellus* Swinh** *See* maize stem borers.

Chiridia semotata **B** *See* stem borers of sugarcane.

Chitin A hardened non-living exoskeleton composed by this substance in insects.

Chrotogonus **spp.** Surface grasshopper of wheat.

Chrysalis During the first ten to fifteen days after, the pupa of silk moth has become completely enclosed active metamorphosis. It becomes changed into a brownish tubular organism known as chrysalis. This finally metamorphoses into the baby insect or the imago. Once the imago is formed, it is ready to escape from the surrounding silken cocoon. This it does by moisturing one end and forcing its way out through the softened silk.

Chyusalis The pupa of insects of the order Lepidoptera (butterflies and moths). *See* pupa.

Circulatory system in insects In cockroach when the terga are carefully removed without any of the underlying tissues, the long mid-dorsal blood vessel is exposed. It is closed behind and is divided in two parts (*i*) the posterior pulsating portion of the heart, and (*ii*) the narrow tubular anterior portion or the aorta. The dorsal blood vessel is the only blood vessel in the body and it lies enclosed in a pericardial sinus or cavity.

Cirphis unipunctata **Haw** *See* paddy cutworm of rice.

Citrus aphids of citrus crops *T. aurantii* Fab., *Toxoptera citricidus* Fab. The dark aphids live on tender leaves and shoots, and suck sap. Spray with 0.2 per cent BHC or 0.05 per cent nicotine sulphate @ 1.2 gallons per tree.

Citrus green bug of citrus crops *Rhynchocoris humeralis* Thunbed. The nymphs and adults of this bug suck juice from fruits causing their premature shedding. Control same as for citrus white flies.

Citrus leaf miner of citrus crops *Phylocnisits citrella stainton.* A caterpillar of a small white moth with fringed wings that feeds inside the leaf tissue and pupate therein; moth lays eggs singly on leaves and shoots. Spray with 0.5 per cent DDT in winter and with 0.25 per cent DDT or 0.1 per cent BHC or 0.25 per cent parathion in other seasons; repeat spraying two to three times @ 1–2 gallons per tree.

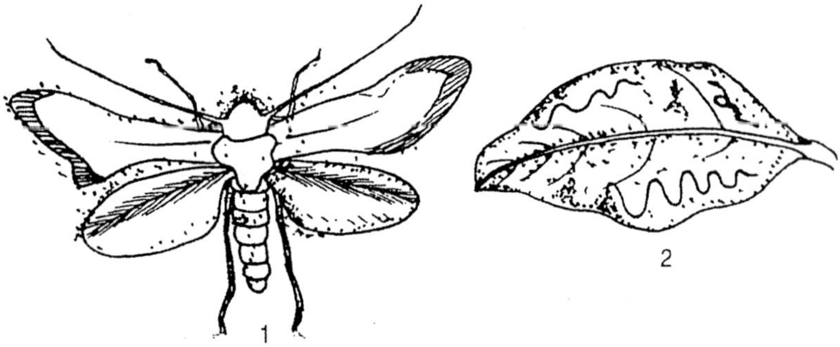

Citrus Leaf Miner Moth
1: Adult; 2: Mined leaf

Citrus psylla of citrus crops *Diaphorina citri* Kuw. A small brown flying insect that sucks sap from new leaves and shoots; its nymph which is flat and yellow, turning brown later also sucks sap; leaves curl and shrivel. Spray with 0.06 per cent nicotine sulphate or 0.02 per cent endrin @ 1–2 gallons per tree.

Citrus white flies *Aleurocanthus spiniferus* Quaintops. *A woglumi* Ashby and *Dialeurodes citri* Ashm. Minute flying insects, allied to aphids and psyllids, that insect leaves and suck sap; they secrete a sticky substance on which a sooty mould develops; attacks tree shed blossoms, and fruit production is scanty; its nymphs behave similarly. Avoid close planting; open out closely planted orchards to air and sunshine; spray with 0.25 per cent BHC or 0.1 per cent HETP or fish oil rosin soap (1 part in 25 parts of water) @ 1–2 gallons per tree.

***Ciania cramèri* West** *See* bagworm of tea.

Claspers *See* legs of butterflies.

Classification of arthropoda The phylum is divided into five classes (*i*) Crustacea, (*ii*) Myriapods, (*iii*) Arachnida, (*iv*) Onychophora, and (*v*) Insecta.

Class insecta The grasshopper, flies, locusts, mosquitoes, lice, butterflies, bees and a host of similar small creatures belong to the class insecta. They are the most abundant widespread of all land animals that can live in dry environment and are the invertebrates that are able to fly.

Climex rotundu The bed bug is somewhat larger than lice and has an oval, dorsoventrally much compressed abdomen and the wings are absent. The common house bed-hug is suspected to be a carrier of typhus, relapsing fever, leprosy and plague. It probably also transmits *Leismania donovani,* which causes Kalazar.

Clinching cutworm *Rhysocia herculea,* can be controlled by two sprays of endosulphan or gunaliphos @ 0.5 per cent 15 days and 35 days after sowing.

***Clisiocampa indica* Walk** *See* tent caterpillar of apple.

Clothes moth There are three species of clothes moth, the larvae of which live on woollen clothing and furs. The naked clothes base-moth. *Tinea Illiela* is the most common. Its wings are of a light colour. The case bearing cloth-moth, *Tinea pellionella,* is the tube building-clothes month or tapestry moth *(Trichophaga tapetzella),* the forewings are black and white. In all these species, the moths are small having a wing spread of about 12 mm. In each case, it is the larvae which does the damage.

Clypeus The ventral portion of frons in insect head.

Cockroach Cockroach of which there are several species afford excellent type of studying the general plan of an insect for they are of convenient size and are also unspecialised. There are two very common species of cockroach in India, *Periplaneta americana* and *Stylopyga orientalis, P. americana* is a native of tropical America and *S. orientalis* of tropical Asia. But these species have become world-wide in distribution on account of the commercial intercourse between different parts of the world through trading ships. *P. americana* is the one usually selected for study. They abound in dark, damp and warm places and feed on various kinds of food with a preference for starchy materials; they are nocturnal in habit and leave an unpleasant odour in their haunts; not injurious to field crops.

Cocoon A protective covering of a larva and pupa.

Cocoon of silk moth Before the fifth moult, the silkworm goes without food for a long time and usually hides itself in a corner. Then the sticky fluid is forced out of the silk glands through a common pore called the spinneret, which is situated on the underlip of the caterpillar. As the sticky material is forced into the air, it forms a long, solid, thin strand of silk. One silkworm can eject about a thousand yards of silk-thread in this manner.

Coddling moth *Laspereysia (Cydia or Carpoeapsa) pomonella,* is a serious insect pest of temperate fruits and causes heavy damage to apples, pears and other fruits also. Repeated sprays with lead arsenate can control this pest.

Cockchafer beetle *Adoretus lasiopygus* and *A. versatus.* Defoliate the grape vine. To control, light traps or stomach spray can be used.

Coelom Body cavity enclosed by mesoderm in Arthropoda. It is greatly reduced.

Colemania sphenareoides **Bot** Deccan wingless grasshopper of jowar *(Sorghum).*

Coleoptera The largest order of insects and possibly the largest order in the animal kingdom, containing the beetles and weevils. Their forewings are modified to form hard leathery elytra, which protect the membranous hind wings and soft abdomen when resting. The head, projected into a snout in weevils, has biting mouth parts. Beetles are found universally in a variety of terrestrial and freshwater habitats. The larvae vary between legless grubs, caterpillar like forms, and predator. Metamorphosis is complete. Many larvae and adults are serious pests *e.g. Anobium* larvae (woodworm) and *Elater* larvae (wireworm).

Others are beneficial *e.g. Coccinella* (ladybird), which eats aphids.

Collembola In this order of Insecta, they have chewing or sucking mouth parts, and antennae with only four segments, lack compound eyes and tracheae, *e.g.* Spring tails.

Colleterial glands These are two in number and asymmetrical. They are very much branched tubular glands. The left colleterial gland is larger and more or less surrounds the right gland.

Colon *See* rectal papillae.

Common oviduct The two short but wide mascular oviducts unite to form a common oviduct which opens into the vagina by a vertical slit, the female gonopore.

Common efferent salivary duct In cockroach, the two glandular ducts unite to form the common glandular duct which opens into the much larger common receptacular duct. Now the duct is known as the common efferent. Now the duct is known as the common efferent salivary duct which opens into the pre-oral cavity at the base of the hypopharynx.

Compound eyes of insects Of all the organs of special sense, the most peculiar of wonderful are the compound eyes of insects. These are a black kidney shaped paired patches on the sides of the head; they are sessile, convex structures, the surface of each of which on being examined with a lens is found to be marked off into a very large number of hexagonal facets.

Conglobate Same as phallic.

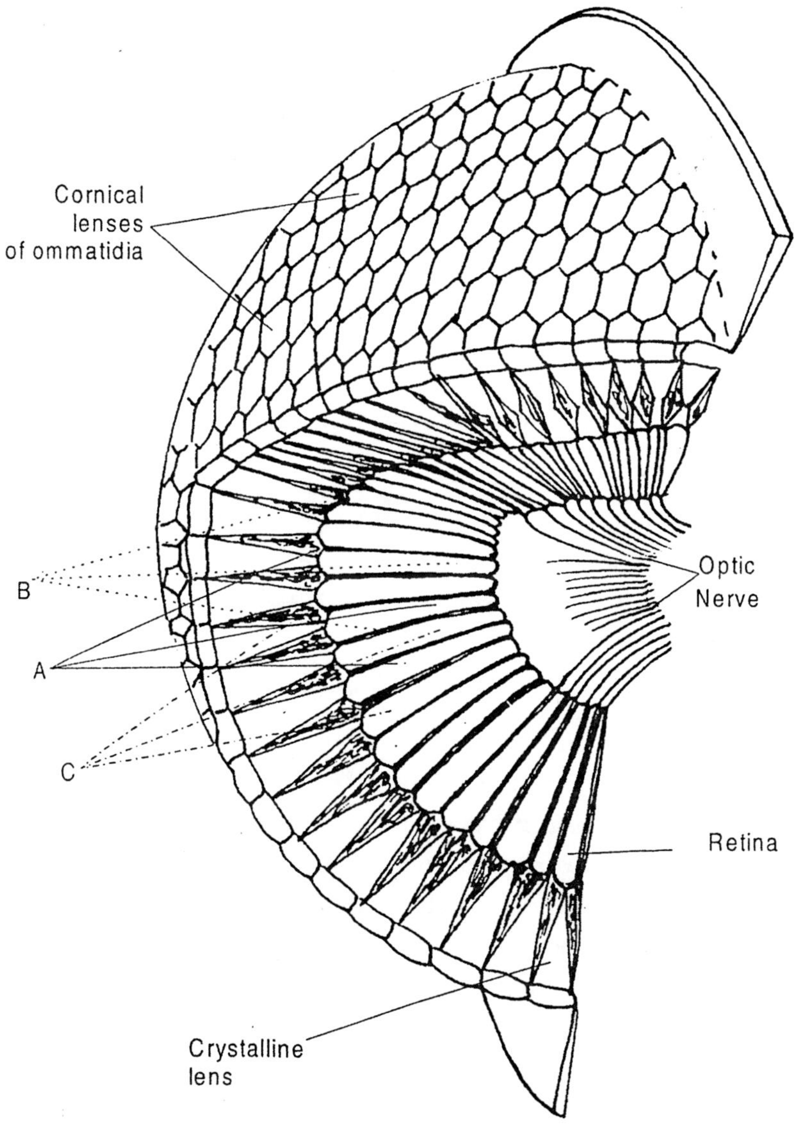

Longitudinal section of a part of the compound eye and compound eye and the rest in the surface view

Contact insecticides Such are the insecticides which act as repellents or may even kill the insect when it comes in contact with it. This kills insects as soon as they come in contact with it. There are various contact sprays, dusts, poisons. The most common cause of death by suffocation by closing the breathing pores or by paralysing the nervous system.

Control of respiration In Terrestrial insects as big amount of water is lost through the tracheal system, it is controlled partly by opening the spiracles just enough to allow sufficient respiration or by the movements of liquid into a out of the terminal parts of the tracheoles.

Control of storage pests These are general in nature and most them are applicable to many pests or group of pests. Preventive measures are more important, though curative measures are also available to suppress or remove infestations.

***Coptops asditicator* F.** A beetle whose grub bores into the stem and kills it. Remove infested plants.

Copulation in cockroach Copulation has been observed only at night from March to September. It takes place about a week after their last moult. The male and female insect touch each other at their hinder ends. The male separates the two gynovalvular plates of female by its titilator and introduces its phallomere into the genital chamber of the female. The spermatophore is now forced out by the mascular contraction of the wall of the ejaculatory duct and it gets directly attached to the wall of spermathecal papilla.

Copulation in housefly The flies come to earth and copulate there only. The male alights on the back of the female by a leap and remains passive during the rest of the operation. The female exerts its ovipositor and inserts it into the genital atrium of the male and then copulation can successfully take place.

Corcyra cephalonica **Stainton** *See* rice moth.

Corneagen cells *See* corneal lens.

Corneal lens in insect eye Externally situated in the transparent, cuticular biconvex corneal lens which is the secretion of the lenticular epidermis cells (the corneagen cells), which lit below it.

Cosmolyce (polyommatus) baeticus **L.** *See* blue butterfly of gram.

Cosmopolites sordidus **Germ** *See* weevil borer of banana.

Cotton bugs (red and dusky) of cotton (*Gossypium* spp.) *Dysdercus cingulatus* Fb., *Oxycarenus Laetus* Kirby. They suck the juice of flower buds and bolls, and taint the lint. Dust with 5 per cent BHC @ 15–20 lbs. per acre.

Cotton jassid of cotton (*Gossypium* spp.) *Empoasca devastans* Dist. A small, greenish leaf-hopper whose nymphs and adults feed on leaves, which turn yellow; plants remain stunted. Spray with rosin soap or 0.2 per cent DDT at 30 to 70 gall per acre or dust with mixture (1 : 1) of sulphur and 5 per cent DDT @ 15–20 lbs. per acre.

Cotton leaf-roller of cotton (*Gossypium* spp.) *Sylepta derogata* F. A small greenish yellow caterpillar that rolls cotton leaves and feeds on them; exotic varieties are more susceptible; the moth is medium-sized and has yellow wings. Pick and destroy rolled leaves. Spray with 0.25 per cent DDT or BHC @ 80-100 gallons per acre. Dust with 5 per cent DDT or BHC @ 15–20 per acre.

Cotton plant louse of cotton (*Gossypium* spp.) *Aphis gossypii* G. Aphids appear on tender portions in colonies; leaves are curled up and plants are blighted. Spray with 0.05 per cent nicotine sulphate or tobacco decoction @ 80–100 gallons per acre.

Cotton spotted bollworm of cotton (*Gossypium* spp.) *Earias fabia* S., *Earias insulana* B. A small, stout, brown caterpillar that bores into top shoots in early stages, and into bolls later on; top shoots droop down and attack bolls shed; it pupates under a dirty brown cocoon on boll surface and occasionally in the soil; emerges as a small green moth. Do not grow lady's finger in off season; remove and destroy attacked shoots and bolls; remove all stumps after harvest. Spray 0.03 per cent endrin @ 80–100 gallons per acre at fortnightly intervals.

Cotton stem borer of cotton (*Gossypium* spp.) *Sphenoptera gossypii* G. A beetle grub that bores into the stem and pupates therein. Uproot and burn withered plants as soon as noticed.

Cotton stem weevil of cotton (*Gossypium* spp.) *Pempheres offinis* F. A small white grub that bores into stem near the base and causes a gall-like swelling; attacked plants lodge with strong wind. Do not grow alternate food plants in off season; remove them, if any.

Cotton white fly of cotton (*Gossypium* spp.) *Bemisia labaci* Lind. Very small flies that feed on leaves on the under surface; affected leaves curl and dry up, and plants remain stunted; also act as vectors of virus disease. Spray with 0.25 per cent BHC or 0.03 per cent endrin @ 80–100 gallons per acre.

Cowpea aphid of cowpea (*Vigna sinensis* Savi) *Aphis craccivora* K. Infests leaves, stem and pods. Spray with 0.05 per cent nicotine sulphate or tobacco decoction.

Coxa *See* legs of cockroach.

Cremaster Same as hooks.

Crickets (*i*) *Gryllus domesticus* Linn, (*ii*) *G. sigillatus* Walker. These are dirty, cream coloured insects, flat and plump with long feelers; inhabit dark, damp and warm places and feed on bread, flour and various other products; sometimes they enter bags and scrape off portions of grain; they are also nocturnal.

Crop In cockroach, the wall of crop when not stretched with food content, are thrown into numerous longitudinal folds which allow its distension. Also *See* mouth-parts of honeybee.

Crustacea This class includes mainly aquatic arthropods in which the exoskeleton is generally strengthened by the deposition of lime salts, example–water fleas and cradles.

Cryptorhychus mangiferae **Fabr** *See* nut weevil of mango.

Crystalline cone *See* vitrelle.

Cucurbita fruit flies *Dacus cucurgitae.* It is active throughout the year except during the cold weather.

Bait with malathion and protein hydrate (1 lb. malathion, 50 per cent W.P. in 1 gallon of water).

Culex pipens It is one of the common mosquitoes. In this insect, there are very efficient mouth-parts which form a very efficient apparatus for piercing and sucking.

Culex pupa It is very different from the larva but unlike the pupal stage in the life history of the housefly, or butterfly, the mosquito pupa is as active as the larva. It resembles, however, the pupa of housefly or butterfly is not taking any food.

Culicidae *See* Animal Science.

Cuticle It is a two layered structure in the three layered body wall of the cockroach. The outer part is too thin and carries pigments and bristles while the inside part is relatively much thickened.

Cut worms *Cirphis loregis, Euxora spiniferi* and *Agrotis yespilon* damage wheat crop. Control measures are same as for army worms.

Cut worms of linseed (*Linum usitatissimum* L.) *Agrotis* spp. Same as for greasy cutworm of gram.

Cut worms of maize *Agrotisispibus* sp. can be controlled by use of Aldicarb 10C (25 C/He) at the time of sowing.

Cyclas formicariusm **F.** See sweet potato weevil of sweet potato.

D

Dactyl Finger.

Dacus ciliatus **loew** The white maggots of these flies of cucurbits bore into the fruit of all cucurbits; it is more active in hot and dry months. Collect and destroy the affected fruits to control.

Dacus diversus Fruit fly of cucurbits. Confined to flowers.

Dairy caterpillar of sannhemp (*Crotalaria juncea* L.) *Utethesia pulchela* L. A hairy caterpillar having red, black and white markings; feeds on leaves and bores into capsules when pods are formed; pupates in leaf fold or in soil; the moth is pale white with red and black spots on upper wings; lays eggs on tender leaves and shoots. Collect egg masses and caterpillar in the early stage and destroy them; dust with 5 per cent BHC @ 15–20 lb. per acre.

Dasyneura lini **barnes** *See* gallmidge of linseed.

Deccan wingless grasshopper of jowar (*Andropogon sorghum* Brst.) *Colemania sphenareoides* Bol. A wingless grasshopper; nature of damage same as by *Hieroglyphus*.

***Deporaus marginatus* pasc** *See* leaf cutting weevil of mango.

Dermaptera This order includes long insects with biting mouth-parts. The fore wings are modified as tegmina. Example–earwigs.

Destructive insects Such are the insects that destroy the products of human industry, agriculture crops, *e.g.* termites and cloth moths.

Destructive household insects Clothe's moth, bed bugs and carpet beetles are the common household insects.

Destructive insects and pests act It was passed by Government of India in 1914 and has been amended from time to time. Quarantine have been imposed by enforcing embargoes, by inspecting at the point of origin and/or destination, and by disinfecting the material.

***Diacrisia obliqua* W.** *See* Bihar hairy caterpillar of sweet potato.

***Dialeurodes citri* Ashm** *See* Citrus white flies of Citrus crops.

Dialeurodes citri* and *D. elongata These two species are serious insect pests of citrus. They are also called Citrus White-fly.

Diamond black moth of crucifers *Plutella maculipennis* C. A small slender, pale green caterpillar that feeds on leaves and makes holes; pupates in cocoon on the leaf; the moth has diamond shaped wings and lays eggs singly. Dust with Pyrodust @ 15–20 lb. per acre. Spray with 0.16 per cent DDT @ 40–60 gallons per acre.

Diaphorina citri **Kuw** *See Citurs psylla* of Citrus crops.

Dichocrocis punctiferalis **G** *See* Castor seed borer of Castor.

Dictyoptera This order of Insects consists of such insects as cockroaches and mantides.

Digestive juice The juice produced by the gastric caeca and the anterior part of the mesontron contains amylase, invertase, lactase, maltase, tryptase and peptase; and thus are carbohydrates, proteins and fats are all digested.

Digestive tract In Arthropods, it is complete, the mount parts are provided with lateral jaws, the mouth parts may be adapted for chewing or for piercing and sucking liquid food.

Diplura It includes small insects with 11 segments of abdomen and many segmented antannae, *e.g. Compodea*.

Diptera The order of insects that contains the flies, characterised by one pair of wings (balancing organs, replace the hind wings). The adults have sucking or piercing mouth-parts and feed on plant juices, decaying organic matter or blood. The legless larvae (maggots) feed on plants, decaying organic matter or are carnivorous. Metamorphosis is complete and the pupa is often protected by a barrel-shaped puparium. Many flies are medically important as the carriers of various diseases; for example the mosquito transmits malaria. Housefly is also cause of many human and cattle diseases.

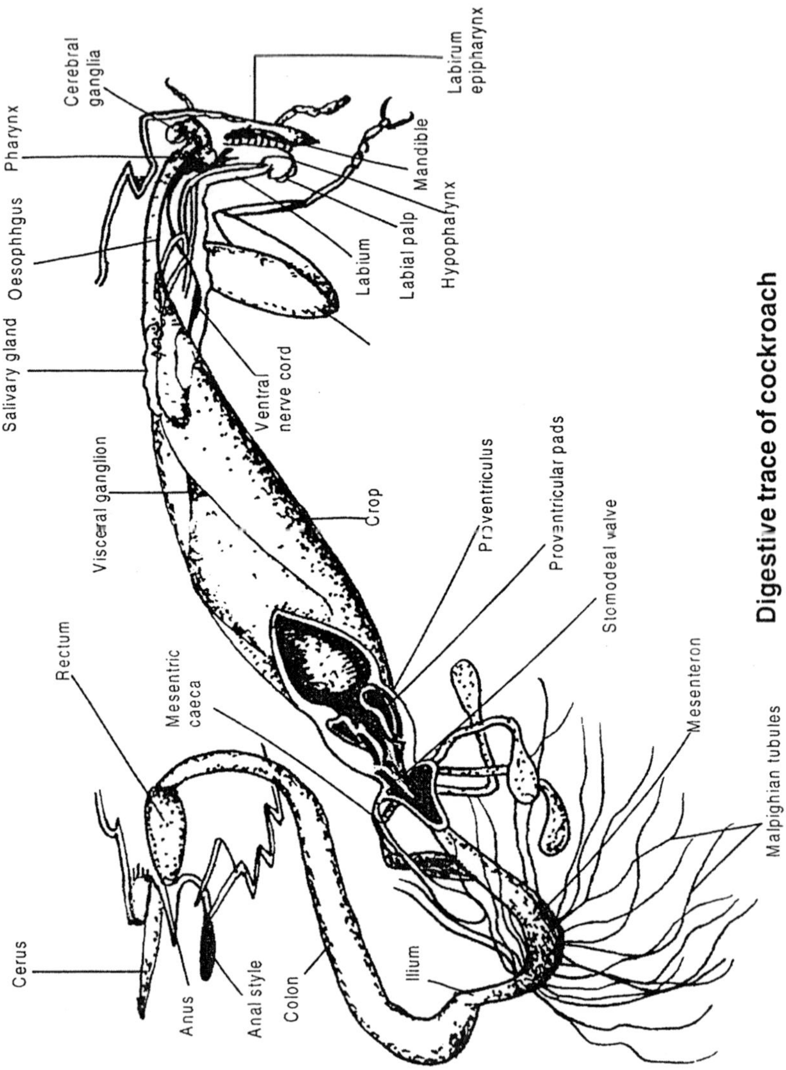

Digestive trace of cockroach

Disinfectants These are chemicals which primarily are used to prevent the occurrence of diseases and insect pests. For their use, some preliminaries have to be achieved of the location, affected animals. These preliminaries are–(1) Through cleaning of the place or locality before application. (2) The phenol coefficient of the disinfectant which indicates the killing strength of a disinfectant as compared to phenol (carbolic acid). (3) Dilution at which the disinfectant is used. (4) The temperature; most disinfectants are much more effective if applied warm or hot. (5) Thoroughness of application and time of exposure.

Distinguishing features of class insecta The grasshoppers, flies, locusts, mosquitoes, lice, butterflies, bees and a host of similar creatures comprise the class Insecta. They are the most abundant and widespread of all land animals that can live in dry environment and are the only invertebrates that are able to fly.

The insects are arthropods in which the body is divided into three regions, the head, thorax and abdomen. The head which consists of six segments bears a single pair of antennae, two compound eyes and the mouth parts. The thorax consists of three segments and is the region which bears three pairs of legs and usually one or two pairs of wings. The abdomen bears eleven or fewer segments and also the genital apertures which are situated near the anus at the posterior end of the body. Excretion is performed by the malpighian tubules attached to the anterior end of the hind gut. Respiration is by branched tracheae (cuticle-lined) that carry oxygen directly to the tissue from paired apertures, the spiracles or

Distinguishing features of the two commonly known genera of mosquitoes

Sl.No.	*Anopheles or Spotted Gnat*	*S.No.*	*Culex or the Common Tropical Gnat*
A.	**Eggs**		
1.	Boat shaped and provides with lateral floats.	1.	Cigar shaped and without floats.
2.	Laid singly.	2.	Found cemented together in floating rafts.
3.	40 to 100 eggs laid at a time.	3.	200 to 400 eggs laid at a time.
4.	Usually found in fresh and clean water.	4.	Usually found in dirty water near human habitations.
B.	**Larva**		
5.	It floats on water surface.	5.	It hangs with head downwards inclined at an angle with water surface.
6.	Has extremely short respiratory siphon.	6.	Has a long conical respiratory siphon.
7.	Dorsal palmate hairs and notched organ present.	7.	Absent.
C.	**Pupa**		
8.	Green and short respiratory trumpets.	8.	Colourless and has longer respiratory.
D.	**Adult**		
9.	Body slender with delicate legs.	9.	Body well-built with stout legs.
10.	Has greyish hairy body.	10.	Greyish body.
11.	Distinctive marketing present; wings spotted.	11.	No distinctive marketing; wings of uniform colour.
12.	Scales present on the abdomen, are closely imbricated.	12.	Abdomen always clothed with dense layer of closely imbricated scales.
13.	Body at an acute angle with the surface on which it rests.	13.	Body more or less parallel with the surface on which it rests.
14.	Palps as long as proboscis.	14.	Palps very short in females.

stigmata, which lie on the sides of the thorax and abdomen.

***Drosicha mangiferae* Green** *See* mango mealy bug of Mango.

Drug store beetle, *Sitodrepa panicea* Linn A brown beetle which bores into turmeric, ginger, coriander and other spices; so also its grub.

Dry fruit moth Same as almond moth. *Ephestia cautella* Wither.

Dusting insecticide on locust *See* locust control measures.

***Dysdercus cingulatus* Fb** *See* cotton bugs (red and dusky) of Cotton.

***D. malignus* Mshll** *See* fruit borers of Apple.

Dysentery in bees This disease chiefly occurs during winter months. Bees in some cases are unable to fly far. They void thin yellowish excreta within the hive, on frames, combs–in fact almost everywhere and also outside on the hive. The disease is caused by bad food and exposure to cold. It may also be due to fermented food in the hive and dampness inside. The disease is of short duration and a few flights by the bees on sunny days will cure it. The stores inside the hive may be examined and unsealed stores, if found fermented, may be removed.

E

Economic importance of housefly The flies are responsible for the death of more people than due to all the wild beasts and reptiles together. Its importance lies in the fact that they are the carriers of millions of bacteria and protozoa among which have been found the germs of such dreaded human diseases as cholera, typhoid, diarrhoea, gonorrhoea, amoebic dysentery, tuberculosis, gangrene, anthrax, leprosy, hookworms, opthalmia, trachoma, bubonic plague, abscemia, tapeworms, roundworms.

Economic importance of mosquito Besides the painful bites inflicted by the female of the mosquitoes, these insects are proved carriers of four distinct human diseases– malaria, yellow fever, dengue and certain forms of filariasis.

Ecosystems *See* Soil Science.

Effect of a mosquito bite After a mosquito has bitten a small area around the puncture whitens, then soon becomes pink and begins to swell and then to itch and burn.

Egg-cocoon *See* hypogynium of cockroach.

Entomology Entomos-Insecta, Logous-discourse. Study of insects.

Ephemeroptera It includes insects in which the adults are found only near fresh water and with aquatic young stages of nymphs. Example–mayflies.

Epicranial suture In an immature cockroach, known as nymph, the vertex is divided by an inverted–Y shaped epicranial suture into two epicranial plates.

Epidermis This layer of the body wall of a cockroach is only single called in thickness, is made up of columnar cells. Its main function is to secrete the cuticle which forms the chitinous exoskeleton. Some of the cells of epidermis produce movable bristles. Such epidermal cells are known as trichogen cells.

Epipharynx The lining of the labrum is known as epipharynx in an insect head.

Equatorial rainforest *See* Plant Science.

Eublema olivacea **M** *See* Leaf caterpillars of Brinjal.

Euproctis lunata **Wlk** *See* Hairy caterpillar of Castor.

Eurentins echinus **D** *See* Sucking insects of Brinjal.

European meal worm *Tenebrio molitor* Linn. The grub of a black beetle, both of which feed on flour, meal, bran, broken grains, refuse, etc., it is usually found in damp, dark, dingy places and godowns. Destroy all termite nests in the vicinity of godowns by treating the soil with dieldrin 0.5 per cent or chlordane 1 per cent. The soil should be raked up and soaked with the chemical.

***Euzophera per licella* R** *See* Brinjal stem borer of Brinjal.

Exarate Designating pupae in which the wings and legs are free and which are therefore capable of limited movement. Compare coarctate, obtect.

Excretory system in insects The principal excretory organs are malpighian tubules and an accessory excretory function is performed by fat body and epidermis. The malpighian tubules are extremely fine, cream coloured or yellow, unbranched blind tubes. They are about 90 in number and are arranged in size to eight bundles, each containing about twelve to fifteen. They lie freely in the haemocoel where they are constantly bathed by haemolymph. Each malpighian tubule is an outgrowth from the beginning of the proctodeum and is lined with a glandular epthelium having a characteristic *brush border.* There are sometimes distinct histological differences in the eptherlium of the proximal and distal portion of a malpighian tubule.

***Exelastis atomosa* W.** *See* Red gram plume moth of Gram.

Eyes of insects Of all the organs of special sense, the most peculiar and wonderful are the compound eyes of the insects. The eyes are a pair of black patches on the sides of the head; they are sessile, convex structures on surface of each of which on being examined with a lens is found to be marked off into a very large number of hexagonal facets. *See* compound eyes of insects.

F

Face fly *Musca autumalis* a close relative of *Musca domestica,* the house fly, is primarily a pest of cattle, particularly of animals in open fields, and away from shade, although it attacks horses and sheep. The face fly does not bite. It feeds on moisture areas around the eyes and muzzle and on blood that oozes from cut, wounds and other such places. Prevention consists in scattering and removing fresh cow manure. Control is difficult but it is recommended Dichlorovosa a 5 per cent B, 3.5 ml, brush on forehead in morning daily if needed.

Facet A division of cornea in the compound eye of an insect or crustacea.

Fat body in insects It fills up greater part of the so called body cavity of cockroach. It consists of white irregular masses or lobes which are everywhere in direct contact with the haemolymph. The fat body receives from the haemolymph the surplus nutritive products of digestion.

Feeding of mosquito When the female mosquito bites, the forked tip of flexible labium is applied to the skin of the victim and acts as a guide and support for the six piercing stylets. All the six stylets enter the skin and the tiny saw-like blades of the first pair of maxillae are worked about as if to saw through the flesh as well as pierce it.

Female gonopore This is situated on the 8th sternum. In a female cockroach, the 8th and 19th sterna of the abdomen get pushed in to form the inner wall of the oothecal and genital chambers.

Female genetalia of insects In case of cockroach, it has two yellow ovaries, one on either side of the median line, lying embedded in the fat body in the 4th to 6th abdominal segment.

Femur The third segment of an insect leg, beween the trochanter tibia.

Fenestra Of an insect, on the inner side of each compound eye, there is a small pale coloured area known as fenestra or ocellar spot. The fenestra are believed to be abortive representative of the simple eye.

Fenestrae These are the vestiges of the simple eyes (ocelli) which occur in many insects.

Filaria bancrofii This is a minute worm which causes Filariasis in human beings. It lives in the blood and lymph. The disease is seldom fatal but is often followed by terrible enlargements of the legs, arms, genital organs or other parts of the body, known as elephantiasis.

Fins of culex The pupa abdomen which is more or less being under the caphalothorax, consists of nine segments, the eight bearing a pair of large fins or paddles which are useful as swimming organs.

Fish oils A large part of the oil sold for poultry consists of a mixture of two or more oils. Fish oils are necessary only when chickens are kept indoors all the time, so that they do not receive the amount of exposure to sunlight necessary for Vitamin D. In India, shark-liver oil is of popular use in poultry rations during summer months when there is a scarcity of green food.

Fish oil soap *See* soap as insecticide.

Flagellum In insects, it is the long many jointed part of an insect antennae.

Flaps In the culex mosquito, at the distal end of the siphon, there are two spiracles which can be completely closed by five terminal flaps.

Flat grain beetle, *Laemophloeus minutus* Oliv The beetle is reddish brown with long feelers; the grub is yellowish white and cigar-shaped; it attacks grain, flour and meal that are in bad condition; the attack usually follows that by other insects as it cannot bore into sound uninjured grain.

Flea The human flea insect are small wingless insects, literally compressed insects which are parasitic only in their adult stage, when they live in the feather or hair and such the blood of birds and mammals. Their larvae are worm-like, collects in cracks under carpets and in the folds of upholstery. The Indian rate flea *Xenoxpsylls cheopis* is the principal agent in transmitting *Bacillus pestis,* the cause of bubonic plague.

Flea beetle of grape-vine (*Vitis vinifera* L.) *Sceleddonta strigicollis* Mots. A small coppery brown beetle that bores into sprouting buds; lays eggs under the bark; the grubs fall down, feed on roots and pupate in the soil. Spray with 0.25 per cent DDT @ 15–200 gallons per acre.

Flight in cockroch The forewings are held at right angles to the body and do not beat, but the hindwings beat in a complicated figure which both supports the body and drive it forward in air.

Flight to death On the emergence of a young queen either a flight to death of swarming takes place, the old queen leaving the hive with a number of workers. Usually a swarm settles on the branch of a tree or some such place. A search is made for a hallow tree or other suitable places. A week after emergence the young queen undertakes her 'nuptial flight' during which she is fertilized by a drone. After this the drone dies but the queen returns to the hive and starts to lay eggs for nearly five years.

Floating raft The number of eggs laid down by a single individual Culex at a time varies from 200 to 300 or more. As the eggs are laid by the female Culex, it arranges them with its hind legs and glues then together side by side making a little 'floating raft'.

Florida red scale A serious pest of Citrus, *Chyrysomphaius aonidium.* Fumigation with hydrocynic acid or powdered calcium cyanide can control the pest.

Fluted sale of citrus crops *Lcerya purchasi* Mask. Also known as cottony cushion scale; the insect is yellow, red or brown in colour and is coated with white or yellowish mealy substance; both adult females and nymphs such sap; attack other trees also. Liberate predaceous beetles *(Rodolia* sp.*)*. Spray with 4 per cent diesel oil emulsion during winter @ 1.2 gallons per tree.

Formicoidae *See* Apocrita.

Frons Below the vertex (cockroach) is a broad squarish plate called frons, which has a ventral clypeus. On the sides of the head, capsule is a gena below each compound eye.

Fruit borers of apple (*Pyrus malus* L.) *Dyscerus fleetcheri* Mshll. Both borers are grubs of weevils, which lay eggs on fruits; the grub makes tunnels in fruits and pupates inside. Collect and destroy weevils, dust of spray fruits with BCH or DDT.

Fruit borers of apple (*Pyrus malus* L.) *Dyscerus fleetcheri* Mshll.

Fruit flies *See* under various fruits.

Fruit fly of custard apple (*Annona squamosa* L.) *Baocera persicae* B. Maggots of this fly bore into fruits. Spray with 0.03 per cent endrin @ 2–4 gallons per tree.

Fruit flies of guava (*Psidium guava* L.) Same as for mango.

Fruit fly of jujube (*Zizyphus jujuba* Mell.) *Carpomyia vesuviana* Costa. A pale brown fly with brownish black spots on neck; maggots feed inside the fruit, and pupate in soil. Remove wild jujube trees; spray with 0.16 per cent DDT @ 2-3 gallons per tree.

Fruit flies of mango (*Mangifera indica* L.) *Strumeta ferrugineus* Fabr and *Strumeta incisus* W., *Strumeta zonatus* Saund, *Dacus ferrugineus* and *D. zonatus.* Reddish brown flies that lay eggs under the skin of fruits; maggots feed on pulp and pupate in soil; fruits fall of premature and show dark, circular punctures; attacks other fruit trees also. Collect and destroy all fallen fruits; spray trees as well as hedges and other vegetation in vicinity with 0.25 per cent DDT.

Fruit flies of plum (*Prunus domestica* L.) Control same as for mango.

Fruit sucking moths of Citrus *Othreis (Ophederis) fullonica* and *O. materna.* Elimination of alternate host trees helps their control. Poisoned molasses with pineapple flavour can also kill the moths.

Fruit sucking moth of citrus crops *Calpe emarginata* Fabr.

Fruit sucking moth of citrus crops *Achoea janata* Linn and *Othreis fullonia* cl. These moths live and breed on certain wild plants growing near citrus orchards; they usually visit orchards at sunset in rainy season and puncture fruits to suck sap; fruits drop off or rot. Destroy wild plants on which the moths breed; spray with 0.16 per cent DDT or BHC two to three times at two to three weeks intervals when the attack is expected. Cover the individual faints with polythene or alkathene bags (about 300 gauge thickness). Bait with lead arsenate and gur mixture (1 part of lead arsenate with 160 parts of gur solution).

G

Galea In an insect mouth, from the inner side of the stripes arises a bipartite structure, consisting of an outer galea and an inner lacinia. The galea is slightly hooded while the lacinia bears numerous stout bristles.

Gall fly of sesame (*Sesamum orientale*) *Aspondylia sesame* F. A mosquito-like fly whose maggot feed on flower buds producing a gall; buds fade and fall. Remove and destroy galls.

Gallmidge of linseed (*Linum usitatissimum* L.) *Dasyneura lini* Barnes. Small greenish white maggots bore into buds and 'young flowers. Spray with 0.2 per cent DDT @ 40–50 gallon per acre. Set up light traps.

Gas insecticides These are such compounds which emit lethal gases to kill the insects and are very useful to prevent insect damage in the stored grain and seeds. The most spectacular use of gas in fruit growing industry is the fumigation of citrus trees with such as hydrocynic-acid-gas to control scale-insects. Powders such as pyrethrum, are generally classed as contact insecticides but also give off gases which kill the insects-pests.

Gastric caeca *See* mesontron.

Gasterophiloidae Bot flies. *See* Animal Science.

Gastro-intestinal nemtodes, worms Internal parasites of cattle include many species which may be found in the stomach, small intestine and/or large intestine including the following: twisted stomach worm, medium stomach worm, minute stomach worm, modular worm, hook worm, cooprids and lungworms. Most of these parasites have similar life cycles, of which the common stomach worm is typical. Infected animals show loss of weight, retarded growth, anaemia, diarrhoea and/or lowered resistance to other diseases. Preventive and control measures include–(1) rotating pastures, (2) segregating calves from mature animals, (3) cross-grazing with cattle and horses, (4) avoiding overstocking or overgrazing pastures since most of the infecting larvae are on the bottom of grass, and (5) keeping feeders and waters sanitary. The general recommended treatment for the removal of adult parasites of cattle which should be administrated in spring and again in winter, following the pasture season consists of: Thiabendazole–as a drench in a bolus, or mixed with the feed–given according to manufacturer's directions (usually 3 to 5 grams per 100 pounds of body weight).

Gastrophilus intestinalis This botfly lays its eggs on the hair of horses which are removed by the tongue. Thus the eggs reach the stomach where on hatching the larvae burrow into the stomach wall and do considerable damage.

GATT (WTO) Agreement It is meant to change the setting of world agriculture through provisions of market access and the consequent economic compulsions in the trade of agricultural commodities.

Gena This is situated on the sides of the head capsule, below each compound eye. To it hinges on the labrum.

Genetalia *See* male gonophore of cockroach. In cockroach, these are contained in the genital chamber, are asymmetrical and consists of three lobes which are arranged around the gonopore in the male insect. According to their position, they are called right, left and ventral phallomere.

Genital chamber *See* female gonopore, the front part of the boat-shaped cavity.

***Ghyptotermes dilatatus* Havil** *See* Para Rubber Termite of Para Rubber.

Girdler Of grapes. A lengieern beetle, *Sthenias grisator.*

Gizzard Same as proventriculus.

Glochidium The larva of a fresh water mussel.

Glossa The bipartite inner portion of the labial pulp in insects. *See* prementum also.

Glossina pulpalis Commonly known as *tsetse* fly carries Trypanosomes of African sleeping sickness.

Gnats These are the insects belonging to family Culicideae. As a rule where the Culicidae are very abundant and troublesome, they are called mosquitoes but where less troublesome, they are called gnats.

***Gnorimoschema (Phthorimaea) glapsigona* M.** *See* Budworm of Brinjal.

Gonopophysis *See* colleterial glands. In the female cockroach, the gonopore is also surrounded by gonapophysis which functions mainly as ovipositor.

Gonopore In insects, the two short but wide oviducts unite to form the common oviduct, which opens into the vagina by a vertical slit, the female gonopore.

Gram caterpillar of gram (*Cicer arietinum* L.) *Heliothis obsoleta* F. A green caterpillar with grey longitudinal streaks; feeds on tender foliage and pods; pupates in soil; the moth is yellowish brown and lays solitary eggs of greenish yellow colour. Dust with 5 per cent BHC or DDT @ 15–20 lb. per acre or spray with 0.16 per cent DDT suspension @ 80 gallons per acre. *Helicovespa armgera*–can be present in the wheat, cotton and gram fields. Fields infested with *Chenopodium* spp. are reported to be most affected.

Gram pod fly of gram (*Cicer arietinum* L.) *Agromyza obtusa* M. A fly that lays minute eggs in pod shells; maggots bore into and damage seeds; sometimes they mine leaves in beginning of season. Clean cultivation; sow resistant varieties; spray with 0.05 per cent nicotine sulphate of 0.25 per cent DDT when maggots are in leaves @ 60–80 gallons per acre.

Grasshopper These are foliage feeders of various crops. *Attractomor phacrenulata, Chrotogonus* spp. are common.

Grasshoppers of sugarcane (*Saccharum officinarum* L.) *Hieroglyphus banian* Fab., *H. nigrorepletus* Bol., *Choroedocus illustris* Wlk. Nymphs and adults feed on the foliage. Dust 5 per cent BHC @ 20–25 lb. per acre.

Greasy cutworm of gram (*Cicer arietinum* L.) and maize (*Zea mays*) *Agrotis ypsilon* Rott. A moth that hides under dried twigs and leaves at day, and lays eggs in clusters on underside of leaves; larvae attack and cut plants at ground level and pupate in soil; moth has dirty brown forewings and creamy hindwings. Mix 5 per cent DDT in soil before sowing @ 20–25 lb. per acre. Dust with 5 per cent DDG, heptachlor, chlordane or aldrin at the base of plants @ 15–20 lb. per acre. Spray the base of the plants with 0.03 per cent aldrin @ 60–80 gallons per acre.

Green bug of citrus crops *Rhynchocoris humeralis* Thunberg. The nymphs and adults of this bug suck juice from fruits causing their premature shedding. Same as for citrus white flies.

Gregariousness Tendency of animals to gather together.

Gragarious locust *See* phases of locusts.

Grey leaf weevil of lady's finger (*Abelmoschus esculentus* L.) *Myllocerus* sp. The adults eat leaf tissue and make holes in leaves. Same as for respective cotton pests.

Grey weevil of soybean (*Glycine max* merr.) *Myllocerus* sp. A small grey weevil that feeds on leaves. Dust with 5 per cent BHC @ 15–20 lb. per acre.

Grey weevil *Myloceros discolor* feeds on leaves and ear-heads of wheat.

Grub The first instar larva of housefly. The cattle grubs are called warbles or heel-fly or gad-flies. These are two distinct species that commonly occur : (1) *Hypoderma bovis* and (2) *H. lineatum*. It injures hides of animals, give shocks to animals. The female deposits about three hundred eggs, usually on the lower legs and particularly just above the hoof (hence the name heel-fly). It is short lived, neither the male nor the female surviving more than about a week. Complete prevention of cattle damage within any given area cannot be obtained unless all cattle grubs throughout the country are exterminated. It needs a nation wide campaign to eradicate the menace, all concerned should cooperate. The recommended insecticides and formula for cattle grub is the same as for other flies.

Gryllus domesticus **Linn.** *G. sigillatus* Walker Crickets.

Gujhia weevil ***Tonyincuss beteceae.*** It is a wheat insect pest. The adults as well as grubs damage newly germinated top by cutting stem at the ground level. Dusting crop with HCH 10 per cent is useful.

Gynatrium *See* hypogynium of cockroach.

H

Haemolypm The heart of cockroach is made of 13 segmentally arranged funnel shaped chambers, each of which is placed in communication with the pericardial sinus or cavity by a pair of laterally placed ostia which permit the entry of colourless blood known as haemolymph. Within the chambers occur volvular folds that allow the blood to move forward and not in the reverse direction. *See* heart of cockroach.

Hairy caterpillar *Euproctis virguncula* is a common pest of many crops.

Hairy caterpillar of castor (*Ricinus communis* L.) *Euproctis lunata* Wlk. A dark reddish brown hairy caterpillar that feeds on leaves and pupates in soil or lead folds. Dust with 5 per cent BHC @ 30–40 lb. per acre.

Hairy catterpillar of drumstick (*Moringa plerygosperma* gaertn.) *Euproctesterote molifera* W. A fairly long, dirty brown caterpillar that feeds on leaves and bark; the moth is yellowish brown and lays eggs on leaves and tender shoots. Collect and destroy caterpillars. In case of severe infestation, dust with 5 per cent BHC @ 1–4 lb. per tree.

Haltere One of a pair of club-shaped structures protruding from the sides of the thorax of flies (Diptera). They are highly modified hind wings and vibrate rapidly during flight to provide information and for the maintenance of balance, in mosquito.

Halteres of housefly A pair of very small drum stick shaped halteres, which are reduced hind wings, lie hidden by scales at the hind borders of the bases of the fore wings. The halteres are provided with sense organs and are believed to be concerned with the maintenance of balance during flight. They are also said to be capable of appreciating sound vibrations.

Hand picking Sometimes it is useful to pick up the damaging insects with hand and destroy them by any possible means.

Harm caused by ticks *See* Animal Science.

Harvest of Chigger mites *See* Animal Science.

Haunch moth of sesame *Achrontia stys.* Controlled by parasite like *Telemonus* sp., *Trichogranoma auatrilicum* and *Anastalus achrontial* who feed on eggs and *Bradiciociol* spp., *Egolithrina autriopionera* and *Apantalis cectial* parasites feed on larvae of *H. styx* of sesame crop.

Haustellum Same as the mediproboscis of the mouth parts of housefly.

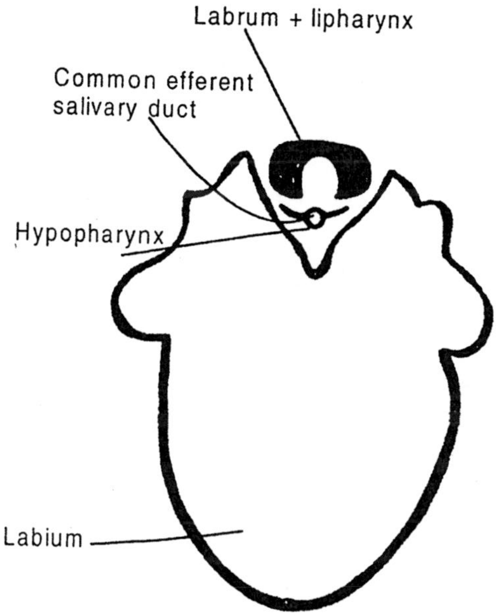

T.S. of haustellum of Musca

Head of housefly Like all other insects, the body of the housefly has three distinct regions–the head, thorax and abdomen; the greater part of its hemispherical head is occupied with a pair of compound eyes while three ocelli or simple eyes lie on the top of the head in the black strip. Each compound eye is composed of 4000 ommatidia which superficially appear as hexagonal facets, as each eye stands out prominently from the head it can catch light from every possible direction. Thus the housefly has much wider field of vision and can see the approach of its enemies without any difficulty.

Head of insects The head consists of six segments and bears a single pair of antennae, two compound eyes and the mouth parts. Connecting the head to the thorax

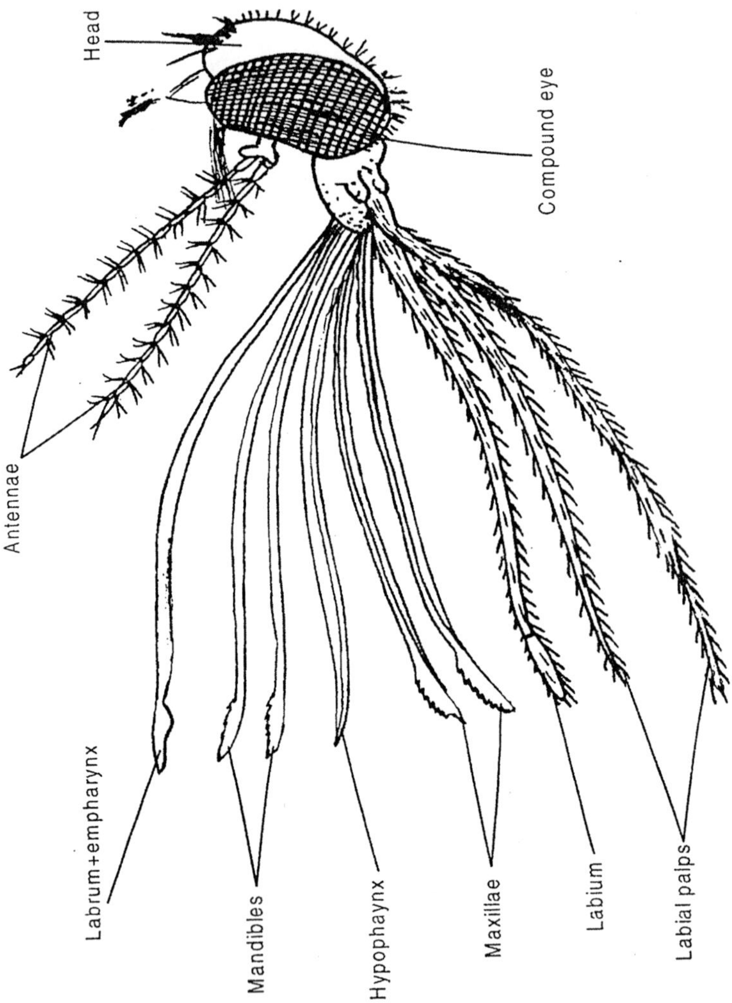

Side View of the Head of Female Anopheles with Parts of the Proboscis Separated

is a slender soft region, sometimes called the neck or cervix.

Heart of insects In cockroach, heart is made of 13 segmentally arranged in funnel shaped chambers, each of which is placed in communication with the pericardial sinus or cavity by a pair of laterally placed ostia which permit the entry of colourless blood known as haemolymph. Within the chambers occur volvular folds that allow the blood to move forward and not in the reverse direction.

Heart of cockroach It is closed behind the terga and is divided into two parts (*i*) the posterior pulsating portion or the heart and (*ii*) the narrow tubular anterior portion or the oarta. The dorsal blood vessel in its body and it lies enclosed in a pericardial sinus or cavity.

Hellebore An important plant product used as poison insecticide. It is made by grinding the roots of the white hellebore plant. It contains alkaloids which are effective both as poison and as contact insecticide. It is harmless to man and can be used on fruit which is almost ripe. It is comparatively expensive and therefore not widely used in horticulture.

Heliothis obsoleta **F** *See* Gram caterpillar of Gram.

Hellula undalis **F** *See* Cabbage borer of Crucifers.

Helopeltis antonii **S** *See* Tea mosquito of Tea.

Hemiptera A very large order of insects, the bugs, characterised by piercing and sucking mouth parts modified into a beak (rostrum). The plant bugs (*e.g.* Aphis) feed on plant sap and are serious pests, both damaging plants and carrying disease. They have uniform for wings and membranous transparent hind wings, which are folded over the back when resting. Wingless parthenogenetic generations are common. The bloodsucking bugs *i.e. Cimex (bed bug)* and the carnivorous water bug *i.e. Gerris (pond sakter),* have fore wings with tough leathery bases, which are folded over the transparent hind wings when resting.

Herse convolvuli **L.** *See* Sweet potato sphinx of Sweet Potato.

Hessian fly A worst enemy of wheat plant. There are many other insects pests of animals and crops and stores which are given in other pages of this book.

Heterusia cingala **Moore** *See* Tea red slugs of Tea.

Hieroglyphus banian Fab *See* Grasshoppers of Sugarcane.

Hieroglyphus banian **F** *See* Rice Grasshoppers of Rice.

Hieroglyphus nigrorepletus **Boised** *See* Maize grasshopper.

Hind gut Same as proctodeum of alimentary canal.

Hippoboscidae Tick flies or louse flies. *See* Animal Science.

Hispa armigera *See* Rice hispa of Rice.

Histogenesis This is the reorganisation of the pupal body of butterfly, giving rise eventually to an adult (or imago form in other insects). *See* pupation.

Histolysis Among insects, most of the larval organs with the exception of the central nervous system are broken down, largely by the action of the phagocytic blood corpuscles into an elemental cream. This breaking down process is called histolysis. The elemental cream forms food for the nourishment of groups of cells are called imaginal discs or buds. It is the extensive breakdown that a larva of butterfly undergoes as a pupa and at this stage neither feeds nor moves.

Holometabolous Insects having a complete metamorphosis.

Holometabolic insects Insects with complete metamorphosis are of this category. It is now considered to be a stage during which the larval body and its internal organs are remodelled to the requirement necessary to adopt them to the need of the adult stage.

Honey bee *Apis mellifera,* are not only useful but also very interesting. This is chiefly on account of highly organised societies which they form. Each swarm or colony or society of bees individuals of three types– (1) Functional female or queen, (2) Functicnal males or drones, and (3) Sterile females or workers.

Honey comb The chief duty of the builder-bee is to construct the waxen comb, which protects the honey, the eggs and developing young. The comb is built in such a way that it consists of double series of hexagonal cells, placed 'back to back' in horizontal rows. The honey and brood comb differ slightly from each other, the cells of the former being filled with honey and capped with wax when they are full while the cells of the latter contain the eggs laid by the queen. The brood cells designed as nurseries for the drones are rather larger than those for the workers, while the few queen cells are enormous vase-shaped structures projecting from the general surface.

Honey sac *See* mouth parts of honey bee.

Hooks *See* pupation in butterfly and also legs of butterflies.

Hornfly Also called cattle fly, stock fly or cow fly. This fly is called *Haematobia irritans,* is about one-half the size of an ordinary housefly. The control of this animal is the as for other flies.

Host *See* Animal Science.

Household insects

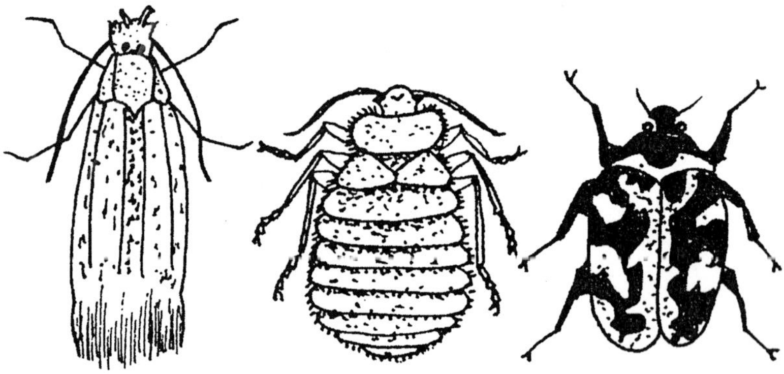

Clothes Moth (Lepidoptera) Bedbug (Hemiptera) Carpet Beetle (Coleoptera)

Household Insects Pests

Housefly ***Musca domestica*** It may not be assumed that any fly caught or seen in the house is the housefly. There are common types of these flies found in our houses, are blue bottles or blow flies. Most of them resemble the common housefly in general structure, habits and life history but none of them are so domesticated or abundant. The housefly can be distinguished from the rest by the presence of four longitudinal streaks over the front part of the thorax and a singular stripe along the back of the abdomen.

The rest of the body is greyish in colour and the underside of the abdomen is mostly yellowish.

Hymolymph *See* pericardial cinus.

Hymoneptera It is very large group of insects, having two pairs of membranous wings and mouth parts used for biting and sucking. Abdomen usually constricted (wasp waist), many form larger social groups. This order of insects contained the bees. A large order of insects containing bees (*e.g. Apis)*, wasps (*e.g. Vespula)*, and ants (*e.g.* Formica). Most are carnivorous, with biting mouth parts, although some (*e.g.* bees) have additional sucking mouth parts. The hind wings are coupled to the larger fore wings by small hooks for a more stable flight, and the thorax is usually jointed to the abdomen by a narrow waist. The female's ovipositor may be modified as a saw, drill or sting. The larvae are either caterpillar-like, feeding on plants, or legless and helpless, being cared for by the adults. Metamorphosis is complete. Although some *Hymenoptera* are solitary; the order also includes social insects, such as honeybees, ants and termites, living in highly organised colonies.

Human itch mite Causes itch in human beings.

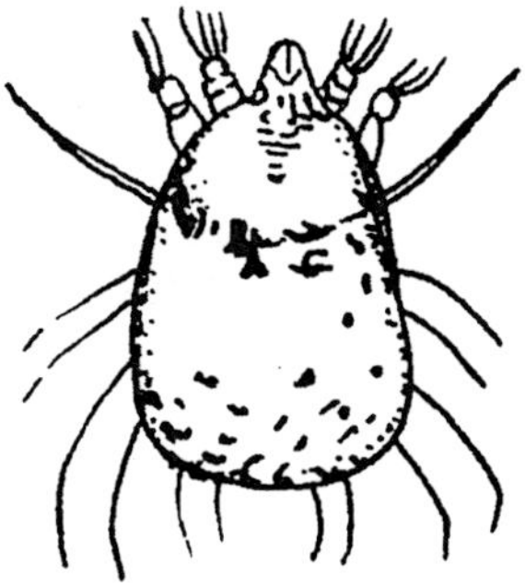

Human Itch Mite (Dermanyssue)

Hypoderma **spp. (warble-flies)** These flies cause great economic loss to the hides and skins industry in the manner to be described presently. The two well-known species of ox warble-flies are *Hypoderma lineatum* and *H. bovis*. The latter has so far been seldom encountered in India, but H. lineatum is widely prevalent in the drier parts of this country. The ox warble-fly deposits its eggs on the hair of legs (especially near the hoofs) and the dewlap. The eggs hatch in about five days and the larvae, after boring through the skin, migrate during the course of about 6 days.

Hypogynium of cockroach The seventh sternum of the female forms a large boat shaped structure called hypogynium. This forms the floor and side walls of a chamber known as gynatrium. The posterior parts of the seventh sternum is subdivided into a pair of gynovalvular plates. The posterior part of gynatrium is known as oothecal chamber, which serves as a mould for the egg-cocoon or ootheca during its formation. The eight and ninth sterna form the roof of the gynatrium.

Hypopharynx In insect mouth, attached to the posterior wall of the preoral cavity and projecting freely into it is the hypopharynx or lingua.

I

Icerya purchasi **Mask** *See* Fluted scale of Citrus Crops.

Idiocerus clypelis **Leth** *See* Mango hoppers of Mango.

Ileum The hind gut has a short, tubular ileum.

Imaginal buds in housefly Same as imaginal discs. These are found distributed throughout the larval body in a situation appropriate for the development of the adult organs.

Imaginal discs in housefly These groups of cells which were clearly set a side and these groups of cells are called imaginal discs.

Imago A baby stage of an insect *e.g.* in housefly and the silk-moth *(plural, imagines).* The final or adult stage in the life cycle of an insect. Unlike the earlier stages, the imago can fly (except in flightless species) and reproduce.

***Inderbela quadrinotata* Walker** *See* Bark eating caterpillar of Mango and citrus crops.

Indian meal moth, *Plodia interpunctella* Hubn A moth with fringed wings that lays eggs on food materials; the caterpillar feeds by preference on flour and other milled products and also on grains; caterpillar stage milled products and also on grains; caterpillar stage lasts for 30 to 35 days and the whole life-cycle about.

Indigenous products for control of insects *See* Animal Science.

Indigo psyllid of indigo (*Indigofera tinctoria* L.) *Arytaina punctipennis* Cr. A small light brown bug with spotted wings that sucks the plant sap; leaves curl up and a knot formed at top of stem. Cut off badly infested shoots; spray with fish-oil rosin soap (1 part in 40 parts of water 0 or 0.06 per cent nicotine sulphate @ 100–150 gallons per acre).

Insecta The largest class of arthropods and the largest in the animal kingdom. Most insects can fly. The body is characteristically divided into a head, thorax and abdomen. The head bears a pair of antennae, compound eyes and simple eyes (ocelli). The mouthparts are modified according to the diet. The thorax bears three pairs of five-jointed legs and typically two pairs of wings. The abdomen is usually limbless. Most insects are terrestrial and respiration is carried out by tracheae with segmentally arranged spiracles. Excretion is by Malpighian tubules. Usually the life history includes metamorphosis but in some metamorphosis is incomplete the larvae (nymphs) resemble the adult and there is no pupal stage. Many

insects are beneficial, being pollinators of flowers and predators of pests; others are harmful, being pests of crops, disease carriers and destroyers of clothes, furniture and buildings. See also mouthparts, Anopura, Aphaniptera, Coleoptera, Dermaptera, Dictyoptera, Diptera, Hemiptera, Hymenoptera, Lepidoptera Odonata.

Insects injurious to plants The Hessian fly is the worst enemy of wheat plant. The flay lays its eggs between leaf bases and the stem, and the larvae leaf bases and the stem and the larvae which hatch out each their way into and through the stalk. The larvae of corn borers attack maize, sorghum and other species of cultivated plants. Cotton boll weevil is the most serious enemy of cotton growers. The larvae which hatch out weevils attack the sweet potato, strawberry, plum and many other fruits and garden vegetables.

Cotton Boll Weevil (Coleptera)

Potato Beetle (Coleptera)

Striped Cucumber Beetle (Coleptera)

Insects Injurious to Plants

Grasshoppers, cricket, scale insects thrips, locusta etc., are also very destructive.

Insecticides Chemicals of plant and chemically originated compounds used for killing and controlling attacking insects of fruit and other crops and animals.

Insects in relation to man Insects are man's chief competitors in the struggle for existence. They eat his crop before he can harvest them and eat them after they are harvested, kill trees, lay waste forests and destroy woodwork eat his clothes, infest his body and distribute horrible disease.

Insect ovaries In cockroach, each ovary consists of eight long beaded tubes known as ovarioles or egg-tubes in which the developing ova are arranged in a linear row in order to development.

Insect pests Some insects simply annoy man such as cockroaches, ants, wasps and nocturnal moths, which are attracted by light. The activities of mosquitoes and housefly are all well-known. The tsetse fly (*Glossina palpalis*) carries trypanosomes of African sleeping sickness. Lice which are small, wingless insects live on the skin of animals and suck their blood.

They are cause of carrying disease germs which harm man and others, like eat away crops, as external parasites of man and domesticated animals, fowls, etc.

Insect pests of barley (*Hordeum vulgare*) Same as for wheat.

Insect pests of cucurbits *Epilachna beetle sp.* Same as for brinjal.

Insect pests of bajra (*Pennisetum typhoideum* Rich) Same as for wheat.

Insect pests of oats (*Avena sàtiva*) Same as for wheat.

Instar The period between moults in an insect larva.

Insectivora An order of small, primitive insectivorous or omnivorous and generally nocturnal mammals *e.g.* Sorex (shrew), Erinaceus (hedgehog) and Talpa (mole). Insectivora have a long tapering snout with sensitive vibrissae and numberous small teeth with pointed cusps for crushing insects. Most have five-clawed digits and locomotion is plantigrade.

Insectivore An animal that feeds mainly on insects. Insectivores are generally small mammals which specially modified teeth *e.g.* the hedgehog, shrew and vole.

Isoptera It includes social insects with very different types of individuals such as sterile soldiers, workers and reproductive forms. The latter are usually winged and the former are wingless. Example–termites.

Instar The stage between successive moults of insects (*e.g.* locusts and aphids) that develop by incomplete metamorphosis. They early instars, in which the insect is small, sexually immature, flightless and often differently marked, are called nymphs. In the later instars, wing buds may appear but these do not become capable of flight until the final moult, at which the imago (adult) emerges. Also *see* larva of housefly.

J

Jack bud weevil of Jack fruit (*Artocarpus integrifolius* Auth.) *Ochyromera artocarpi* M. Grubs bore into the tender buds and fruits. Spray with 0.25 per cent DDT @ 0.25 per cent per tree; destroy infested and fallen buds and fruits.

Jassids, empoasca kerri var motti Pruthi Green soft bodies insects that suck sap from leaves and other tender parts. *Typhlocyba maculifrons, Zygidia malaniesis*– damage the wheat crop sometimes very seriously. Spray crop with 0.16 per cent DDT @ 60–80 gallons per acre.

Jujube caterpillar *Nephopteryx bigraphella* is a serious insect pest of Litchi tree also. It feeds on leaf and flower, and the tender fruits and leaves causing considerable damage to the crop. Can be controlled by a spray of Calcium arsenate.

Jute semilooper of jute (*Corchorus cappsularis* L.) *Anomis sabulifera* Guen. A long greenish caterpillar that feeds on leaves of top shoots; growth of plants checked; larva forms a cocoon and pupates in the soil. Shake off the larvae in water containing kerosene; spray with 0.2 per cent BHC at 40 to 50 gallons per acre or dust with 5 per cent BHC at 15 to 20 lb. per acre.

K

Khapra beetle, *Trogoderma granarium* Everts A brown oval beetle whose grubs damages grains of almost all cereals, millets and pulses; infestation restricted to top layer, but in severe cases the entire stock may be involved; the grub is brownish white with bundles of reddish brown hair all round the body.

Kinds of locusts These are eight main species of locust, of which three *viz.* Bombay locust, Migratory locust and Desert locust are found in India. The first two species live and breed in the country and are found scattered all over western India from Rajasthan in the north to Mysore and Madras in the south. They rarely form swarms. They mainly feed on plants of the grass family and occasionally, when they become gregarious.

L

Labellae *See* mouthparts of housefly.

Labium In the insect mouth, it is a lower lip which is a median structure attached just behind the masillae.

Labial palp *See* prementum.

Labrum This is erely a plate, hinged on to the lower end of clypeus forms the roof of the preoral cavity, overlaying the mandibles in cockroach.

Lacinia *See* galea.

Lac insect of jujube (*Zizyphus jujuba* mell.) *Laccifer lacca ker.* If lac is not wanted, prune and bum infested branches.

***Laccifer lacca* Ker** *See* lac insect of jujube.

***Laemophloeus minutus* Oliv** *See* flat grain beetle.

Lancet processes *See* mouthparts of mosquito.

***Laphygma exigua* Hb.** Same as for gram.

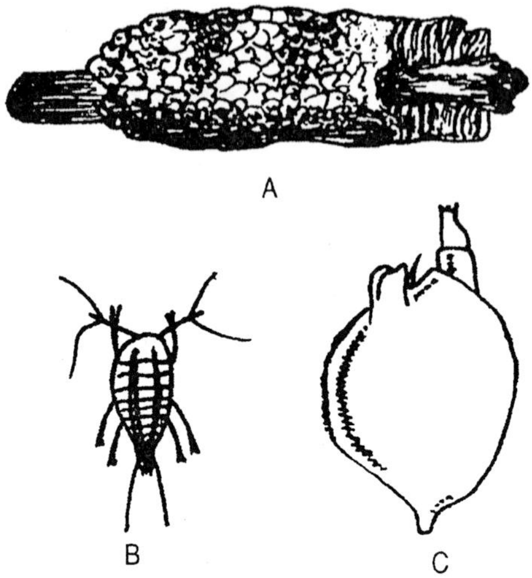

The Lac Insect

Laphygna exigua **Hb.** *See* lucerne caterpillar of linseed.

Larva The young of an animal which undergoes metamorphosis after it has hatched from the egg. It is usually an actively feeding stage characterised by rapid growth.

Larva of culex The eggs develop into tiny young, which are quite unlike the winged mosquito. The young ones leave the eggs by the lower end when hatching takes place and thus drive straight out of the eggs in the water. This stage is called the larva, also known as the wriggler.

Larva of housefly Usually after 8 to 24 hours of deposition the eggs hatch. A minute split occurs at the anterior end of the dorsal side and continues posteriorly enabling the larva or maggot to crop creep out. The larva is encased in an inexpansible case of cuticle, in order to make room for increase in size, the outer skin must be thrown off. This process is known as moulting. The period between moults is known as stadium while the form of the larva between moults is an instar. Upon hatching from the egg emerges the larva which is the first instar. The larva of the housefly moults twice and thus there are three instars.

***Lasioderma serricorne* Fabr** *See* Tobacco borer beetle.

***Latheticus aryae* Waterh** *See* Long-headed flour beetle.

Leaf and fruit eating beetles of apple (*Pyrus malus* L.) *A. lineatopennis* B1.

Leaf and fruit eating beetles of apple (*Pyrus malus* L.) *Adoretus duvauceli* B1.

Leaf and fruit eating beetles of apple (*Pyrus malu* L.) *Melolontha nepalensis* B1. Beetles lay eggs in the soil during the rainy season; grubs feed on vegetation till next summer; beetles come out in June and feed on leaves and fruits at night. Collect beetles by shaking trees and destroy; spray with 0.25 per cent lead arsenate or 0.15 per cent to 0.25 per cent DDG @ 2 to 5 gallons per tree.

Leaf and fruit eating beetles of apple (*Pyrus malus* L.) *Anomala polita* B1.

Leaf and pod caterpillar of sesame (*Sesamum orientale*) *Antigastra catalaunalis*. A small brown moth whose pale green caterpillar feed on leaves and bore into shoots and fruits; caterpillar makes a web with top leaves by folding them, and pupates in a greenish white cocoon. Affected shoots and leaves may be hand-picked and destroyed. Dust with 5 per cent BHC @ 15 to 20 lb. per acre. Spray with 0.25 per cent DDT or BHC @ 40 to 50 gallons per acre.

Leaf beetles of sweet potato (*Ipomoea batalas* Lam.) *Aspidomorpha miliaris* F. Damage is more or less similar to caterpillars mentioned above. Spray with 0.16 per cent DDT @ 60–80 gallons per acre.

Leaf beetles of sweet potato (*Ipomoea batalas* Lam.) *Chiridia semotata* B, *Metriona circumdata* H and *Oncocephata tugerculata* O.

Leaf caterpillar of brinjal *Phyctia clientella* Z. A small greenish caterpillar and *Psara bipunctalis* F.

Leaf caterpillars of brinjal (*Solanum melongena* L.) *Eublema olivacea* M. A purple brown caterpillar that rolls leaves and feeds on them; the moth is olive green. Remove rolled leaves and destroy caterpillars. Spray with 0.16 per cent DDT @ 60 to 80 gallons per acre.

Leaf curling aphid of peach (*Prunus persica* Batsch) *Anuraphis helichrysi* Kalt. It sucks juice from leaves which get curled and twisted and turn dull reddish brown. Spray with 0.5 per cent DDT in winter; remove golden rod, an alternate host plant, from the vicinity.

Leaf curling mite of litchi (*Litchi chinensis* Sonn.) *Eriphyes* sp. Mites attack leaves which curl and acquire a thick growth of minute brownish hairs on the underside in which they leave and feed. Remove and destroy infested and fallen leaves; spray with lime sulphur wash (1:2:16) or 0.025 per cent parathion @ 2–5 gallons per tree.

Leaf curling weevil of mango (*Mangifera indica* L.) *Deporaus marginatus* Pasc. A small grey brown weevil with shining, black fore wings and a long, brown snout that cuts off new leaves near their base; lay eggs in cut leaves in which the grubs feed and later pupate in the ground. Collect and burn cut leaves; spray with 0.2 per cent DDT in the beginning of monsoon.

Leaf eaters of colocasia (*Colocasia antiquorum* Schott) *Monolepta signata* O and *Prodenia litura* F. A white spotted flea beetle both feed on leaves. Dust with 5 per cent BHC @ 15 to 20 lb. per acre.

Leaf eating beetle of brinjal (*Solanum melongena* L.) *Epilachna* spp. The young ones feed on leaves and other tender parts. Destroy insects by hand picking. Spray with 0.25 per cent DDT @ 60–80 gallons per acre. Dust with 5 per cent BHC or calcium arsenate mixed with lime (1:4) @ 15–20 lb. per acre.

Leaf gall midges of mango (*Mangifera indica L.)* *Procontarinia matteiana* Kieff and others. The larvae of these insects feed on leaves and produce small galls in spring and summer. Collect and destroy affected leaves.

Leaf hopper bug of lady's finger (*Abelmoschus esculentus* L.) *Empoasca d evastans.* Spotted boll worm of cotton, leaf roller and cotton semi-looper also attack lady's finger. Same as for respective cotton pests.

Leaf hopper of grapes *Empoasca dedipiens.*

Leaf hopper of sugarcane (*Saccharum officinarum* L.) *Pyrilla perpusilla* Wlk. A soft bodied, light brown, sucking insect on leaves; nymphs and adults cause considerable damage, especially when in epidemic form; pale white or light bluish eggs are laid in cluster on the underside of leaves, covered with a cretaceous, thready material. Destroy eggs early; collect and destroy nymphs and adults; spray with 0.25 per cent BHC, 0.02 per cent endrin at 100–150 gallons per acre, or dust with 5 per cent BHC at 30 to 40 lb. per acre.

Leaf miner of groundnut (*Arachis hypogaea* L.) *Stomopteryx nerteria* M. A small, dark-headed, greenish caterpillar that folds leaves and feeds on them; the moth is small and dark brown. Dust with 5 per cent DDT @ 20–25 lb. per acre. Trap moths with light in early stages of the chep.

Leaf miner of soybean (*Glycine max* Merr.) *Stomopterynerteria* M. *See* under groundnut.

Leaf miner of citrus and fruit sucking moth of citrus *Phyllocnistis citerilla.* Control by hand picking. *Othrois fullonica*–bit with poisoned mollases flavoured with pineapple essence can kill the moth.

Leaf roller of maize (*Zea mays* L.) *Marasmia t rapezalis* Guen. A greenish larva that rolls and feeds on leaves of various gramineae. Hand pick rolled leaves and destroy larvae. Spray with 0.25 per cent DDT @ 60–80 gallons per acre.

***Lecanium viride* G.** *See* Scale insects of Coffee.

Legs of butterflies Each thoracic segment bears a pair of jointed 'true legs' with claws at the end. These correspond to the legs of the imago. The third to the sixth abdominal segments bear unjointed fleshy prolegs. Each of them bears a row of tiny hooks at the end. Such legs are present only in the caterpillar and not in the imago or the adult. The last segment of the abdomen bears a pair of clappers similar to the prolegs.

Legs of cockroach There are three pairs of thoracic walking legs known as prolegs, mesolegs and metalegs. They are more or less similar in structure. Each of these consists of a linear series of a constant number of joints with five distinct parts. Each leg is attached to the body between the notum and the sternum by a small stout and flattened segment, the coxa. It is followed by a small trochanter which if freely movable on the coxa and is fixed on to the femur which follows. The femur is long and cylindrical and largely moves only along with the trochanter as if it were a part of the latter. The tibia is slightly longer but more slender than the femur. The tarsus is made of five movable parts of tarsomeres, all of which have fine hairs. The terminal tarsomere is known as pretarsus. It ends in two lateral sharp claws and a small median, hairy porous, pad, the aerolium which is an adhesive organ. The lower sides of the tarsomeres also have small, adhesive pads known as plantulae. The arolium and the plantulae enable the insect to walk upon smooth and verticle surfaces and even on window panes, ceilings and other similar places.

Legs of honeybee The wax is passed back between the segments in the form of plates, which are removed from the abdomen by spines borne on first joints of the tarsus of each hind leg and transferred to the mandibles where it is mixed with saliva and then moulded and fashioned into familiar hexagonal cells. The hind legs are considerably modified; both the tibia and the first tarsal joint are very large; the inner surface of the latter carries the pollen brush composed of nine rows of stiff bristles which remove pollen grains from various parts of the body. When the brushes are full the hind legs are crossed, the pollen is transferred by other bristles to the concave outer surfaces of the tibia of the opposite leg. Here it is retained by the large curved hairs on the margin of the hollow which forms the pollen-basket.

Legs of housefly The three pairs of walking legs are articulated from the ventral side of the thorax. Each walking leg consists of the usual five regions–the coxa, trochanter, femur, tibia and the tarsus. The tarsus is five jointed and terminates in a pair of curved claws. Between the claws are a pair of well-developed glandular pads, the pulvillae, which bear innumerable hollow hairs through which a sticky substance exudes. It is sticky substance which enables the fly to walk upside down and also to climb smooth surfaces. All

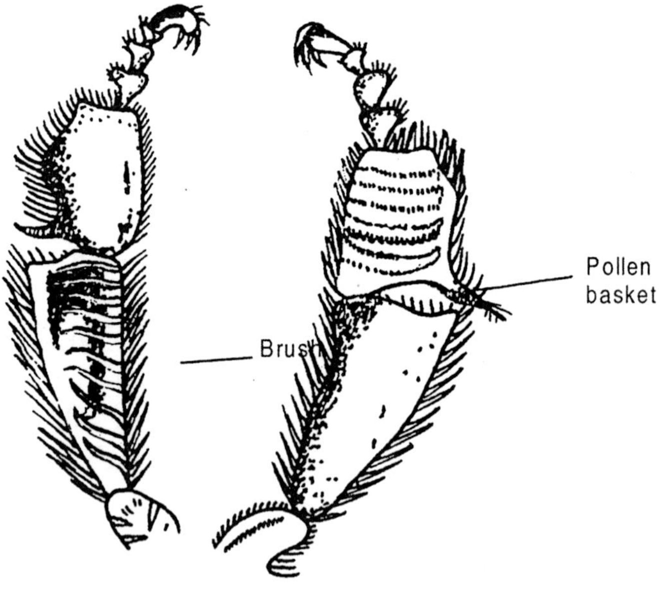

Hind Legs of Worker Bee

the legs are rather thickly covered with coarse hairs of bristles and with a mat of fine, short hairs. One some of the segments used as a sort of comb with which the fly is usually seen combing off the dirt from its body.

Legs of mosquito Each of the three segments of the thorax bears a pair of jointed legs. The three pairs of delicate legs are attached close together to the apex of the thoracic pyramid. In its segmentation each leg resembles that of the cockroach except that coxa is much shorter and tarsus much longer.

Lemon butterfly of citrus A serious insect pest eating the foliage. *Papilio demoleus, P. polytes* and *P. polymnester.* For control, hand picking of the caterpillars, or spray with lead arsenate, repeat at intervals of few days.

Also swallow-tail butterfly that lays eggs on tender leaves; caterpillars which is brownish black and white when young, turning green with brown bands later; feeds on leaves; pupates on the plant remaining suspended. Spray with 0.025 per cent DDG @ 1–2 gallons per tree.

Lepidoptra It includes often large insects with two pairs of membranous wings which are covered with overlapping scales. The mouth parts have elongated galeae with which they are able to suck up the nectar from flowers and have a lengthy metamorphosis. This group includes a hundred thousand species. Example butterflies and moths.

Lepidoptera A larger order of insects containing the butterflies and moths, characterised by a covering scales, often brightly coloured, over their wings and bodies. Mandibles are usually absent and the maxillae form a tube (proboscis) for sucking nectar or fruit juices. The wings are coupled together in flight. The larvae (caterpillars) are mostly herbivores; some are serious plant pests. Metamorphosis is complete, with a pupal stage (the chrysalis). Butterflies are diurnal, have slim bodies and clubbed antennae, and rest with the wings bed antennae, and rest with the wings folded over the back. Moths are mostly nocturnal, never have clubbed antennae, and rest with the wings in various positions.

***Leptocorsa varicornis* Fabr** *See* Paddy bug of Rice.

Lesser grain borer (*Rhizopertha dominica* Fabr) A dirty white grub, having tiny hairs on its body and a curved abdomen, that bores into grains of cereals, millets and pulses; feeds also on cereal products; the adult is a cylindrical beetle of dark brown colour with a roughened body surface; eggs are laid on grain, grain bags and godown walls and crevices; grub stage lasts for 44 days and five generations occur in a year. Immediately before storing, the godown should be thoroughly cleaned and all sweepings burnt and not thrown away; spray the surface of the walls, ceilings, etc. with 0.5 per cent lindane or malathion @ 0.5 litre per 100 sq. metre or fog the godown, warehouse etc. So as to deposit a film of lindane or malathion @ 275 mg per sq. metre; fumigate with methyl bromide (16 mg per cu. metre) for 24 hours or with EDCT mixture (400 gm. per cu. Metre) for 36 hours.

***Leucinodes arbonalis* S.** *See* Brinjal shoot and fruit borer of Brinjal.

Lice These are small wingless insects; live on the skin of animals and such their blood. These species infest man also. The louse is a small flattened wingless insect parasite. There are several species. Hog lice are the largest species, may range upto fourth inch in length. The two main types are the sucking type and biting type. Most species of lice are specific for a particular class of animals. Cattle harbour a number of species of lice including the long nosed and short nosed sucking types of lice and the little red biting ice. Cattle lice will not host any other animal. They attach their eggs or nits to the hair near the skin where they hatch in about in two weeks. For control, Ciodrin chemical's spray is recommended on the body of the affected animal.

Liver fluke *Fasciola hepatica* is a flattened leaf-like brown worm, usually about an inch long. It affects cattle, sheep, goat and other animals. Flukes reproduce by means of eggs which after passing from the host, hatch into the embryos equipped with cilia that enable them to move about. Upon encountering certain types of snails, they penetrate into its body and develop into cercariae (flukes in the larval stage), which leave. the snails and become encysted on the nearby vegetation. The encysted cercariae ingested by animals during grazing. The fluke is liberated from the cyst, penetrates the intestinal wall, migrates about the abdominal where maturity is attained two to three months after infestation. Drainage or avoiding wet pastures, cleaning the pastures of all snail population (by using 6 lbs of copper sulphate per acre of pasture land and treating ponds and sloughs with one part of copper

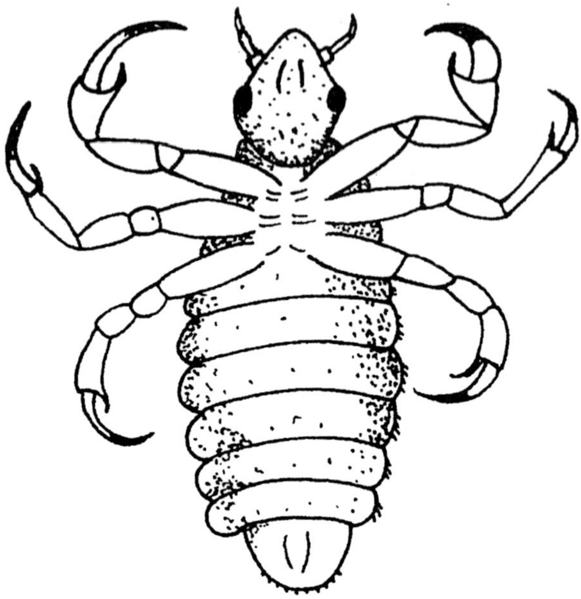

Louce

sulphate to 5000,000 parts of water) can control the menace.

Life history of culex mosquito Although a female mosquito can live on a purely vegetarian diet, it appears that a meal of blood must be taken before it can lay eggs. After mating, the female makes its way to water in a ditch, tank, pond, swamp etc. Still water seems to be necessary and the water must be of no great depth. Egg laying is believed to take place in the early hours of morning.

Life history of housefly About four days after copulation, the female fly starts laying eggs on any kind of decaying and fermenting material. The eggs are laid down at any time during the breeding season which lasts from March to October in India. It lays eggs in the manure of horse but where there is no horse manure it will use, cow, sheep or chicken manure or decaying fruits, flesh or vegetables, ordinary garbage or any mass of decaying organic matter. Each egg is shining or pearly white in colour, cylindrical oval in shape and measures about 1 mm in length. *See* larva of housefly.

Ligula *See* mouthparts of honeybee.

Lime sulphur An important sulphur insecticide. This is also a powerful fungicide. The self boiled form is milder and is seldom used as an insecticide. It is complex mixture of lime sulphur and water, it owes its effectiveness to the presence of polysulphides of calcium, which act as reducing agents breaking down into calcium thiosulphate and finally calcium sulphate. Lime sulphur even when diluted for spraying purpose can damage tender stems. Sprayer

Lime-Sulphur Dilution Table

Density (Degree Raume) Give	*Gallons of Water to be Added to One Gallon of Added Concentration Solution*	
	Sp. gr. 1.010	*Sp. gr. 1.005*
28	22	46
29	23	48
30	24	50
31	26	52
32	27	54
33	28	57
34	29	59
35	30	61
36	31	63

often protect their faces with cold cream and try to avoid getting the spray into their eyes and on the face. It enters the eyes and pains badly for a few seconds. The concentrated material may be handled carefully.

***Linda nigroscutata* Fairm** *See* stem borer of Apple.

***Lineatopennis* Bl.** *See* Leaf and fruit eating beetles of Apple.

Lingua *See* hypopharynx.

***Lipaphis erysimi* (Davis)** *See* Mustard aphid of Crucifers.

***Lipaphis (Rhopalosiphum) erysini* Davis** *See* Mustard aphid of Mustard.

Locomotion of cockroach It normally walk or runs and rarely flies. During walking, three legs are used at a time and these remain in contact with the ground. On one side of the first leg pushes while the third pushes but on the opposite side the second leg supports the body. Meanwhile, the other three legs-two of one side and one of the opposite side-are being moved forward to repeat the process.

Locusts Locusts-invasions cause immense damage to crops and trees, sometimes resulting in severe famines and considerable loss of livestock. Locusts are closely allied to grasshoppers but differ markedly in certain respects. Both are anatomically very similar, feed on vegetation, lay eggs in soil and pass through several instars before they become adults. Where grasshoppers feed on specific crops only, locusts devour almost all plants and vegetation, and even feed upon dead and dying locusts.

Locust control measures There are two main methods of locust control *viz.* (1) use of poison baits, and (2) dusting and spraying with insecticides. Poison baiting though cheap, is not very practical; dusting and spraying have been found to be very effective. In the egg stage, ploughing or flooding the land has been practised in the past. The method is not very effective as it destroys only about 30 per cent of the eggs and renders subsequent control measures by insecticides difficult. Spraying the egg infested land with aldrin is very effective as it leaves a poisonous film on the surface with which the emerging hoppers come in contact and die.

Locust control at international and national levels A locust is an international pest, control measures are effective only when they are properly organised at national and international levels. Since 1951, the F.A.O. has taken active interest in desert locust control by periodically appraising the overall situation, recommending suitable control techniques and organisational patters, and by coordinating and contributing to control campaigns in certain key areas. More recently it has been organising international campaigns annually in the Arabian peninsula with the assistance of a number of countries.

Locust control by dusting or spraying Once the hoppers have emerged, they can be effectively controlled by dusting or spraying or both, as the hopper stage is the most vulnerable. For dusting, BHC is the most effective insecticide. For young hoppers, a 3 to 5 per cent dust is sufficient to kill them; for advanced stage, a 5 to 7 per cent dust is necessary, and adult hoppers need 10 per cent BHC. Dusting or spraying may be done on the hoppers directly or it may be done on strips of ground, two to three yards in width, in the path of the advancing hoppers. Another method is to dig trenches 1 per cent to 2 per cent feet deep and one to two feet wide depending on the size of hoppers, in front of the moving hoppers to which they may be driven and buried alive. Dusting or burning of bushes in which hoppers rest can also be adopted.

Long headed flour beetle, (*Latheticus aryae* Waterh) A pale yellow beetle that infests flours and meals of cereals and dried vegetable matter in storage; the grub which is yellow with numerous hairs on body also feeds on flours and meals. If the grain is infested or gets infested later or, fumigations may be carried out with ED/CT at 25 lb/1000 c. ft. for 72 hours, or methyl bromide at 1 lb/1000 c. ft for 24 hours of Phostoxin tablets at 2.5 tablets per ton for five days. Care should be taken that the moisture content of the commodity is not more than 10 per cent. Only commodities should be treated with a fumigant that do not react with oil.

Lousiness *See* Animal Science.

Lucerne caterpillar of gram (*Cicer arietinum* L.) and linseed (*Linum usitatissimum* L.) *Laphygana exigua* Hb. A dark brown moth, medium sized; lays eggs on lower portion of young plants; caterpillars cut young plants at day time; pupation takes place in soil. Collect and destroy eggs and caterpillars, dust or spray as for gram caterpillar.

Lungworm *Dictiocolus viviparus* is a white thread-like worm 1 per cent to 3 inches long found in the trachies and branchi of cattle especially calves. For control–(1)

Practise rigid antibiotics, (2) Where practical segregate calves from other animals, (3) Keep calves on a good ration, (4) Do not spread infested manure on pastures, (5) Utilise dry pasture if possible. Diethylcarbamazine is effective against lungworm when given as a single intramascular dose of 20 mg/lb or preferably 3 doses of 10 mg/lb. Good care and feeding will help.

Lycaenid butterflies The larvae feed on flower buds and petals. Same as for blue butterfly on gram.

M

Macropes excavatus **Dist** *See* Black bug of Sugarcane.

Macrosiphum **sp.** *See* Safflower aphid of Safflower.

Maggot The first instar of housefly.

Maize stem borer of maize (*Zea mays* L.) *Chilo zonellus* Swinh. Dirty brown moth; nocturnal in habit; lays oval eggs in clusters on underside of leaves; caterpillars feed on leaves and bore into stem and cobs; attacked plants remain stunted or die. Plough up and destroy stubbles after harvest. Spray three times at fortnightly intervals starting when the crop is about a month old with 0.32 per cent DDT @ 40–80 gallons per acre.

Maize (*Zea mays* L.) grass hopper *Hieroglyphus nigrorepletus* Boised. A grasshopper that lays eggs in soil three to four inches deep; hoppers feed on crop right up to adult stage; adults are short-winged and fly short distance only. Destroy eggs by ploughing deep after harvest; dust with 5 to 10 per cent BHC at 15 to 20 lb. per acre.

Male genetalia of insects It has two testes, each of which occupies dorsolateral position and extends from the third to the sixth segment in cockroach. Each testis consists of 30–40 small rounded vesicles in three or four groups but arranged in a longitudinal series.

Male gonopore The ejaculatory duct in insect (cockroach) opens to the outside by the male gonopore which lies close to the ventral phallomere.

Male gonopore of cockroach It is placed below the anus and is surrounded by three asymmetrical lobes, which together form the genetalia. According to their position, they are known the right and the left and ventral phallomeres. They are said to be part of ninth segment. In the female cockroach, the gonopore is also surrounded by gonopophyses which function mainly as ovipositor.

Mallophaga The insects of this group are all parasites either on birds or mammals. They have biting mouth parts and lack metamorphosis. They are very minute. Biting lice *Trichodectes canis* (which acts as an intermediate host of tapeworm *Dipylidium canimum*) and *Heterodoxus spinioerum* on dog, Damalinia (Bovicola) bovis on cattle, *D. equi* on horse, *D. ovis* on sheep, *D. caprae* on goat, *Haematomyzus elephantis* on elephant and *Felocola subrostratus* on cat. The lice infesting poultry are the body louse (*Eomenacanthus strammeus*) occurring especially around the vent and on the breast and thighs. The large body louse (*Gonoides gigas*), the shaft louse (*Menopon gallinae*) on body feathers, the head-louse (*Cuclotogaster heterographus*) on the skin and feathers of head and neck, the wing louse (*Lipcurus caponis*) found on underside of large feathers, and the fluff-louse (*Goniocotes gallinae*) occurring at the fluff at the base of feathers. *Anaticola crossicornis* is found on ducks and *Columbicoa* on pigeons.

Malpighian tubules These are inserted at the anterior end and are yellow, silk-fine, blindly tubular and arranged in six groups, each group containing about 15 tubules. These are the principal excretory organs and an accessory excretory function is performed by fat bodies and epidermis. Each malpighian tubule is an outgrowth from the beginning of the proctoderm and is lined with a glandular epithelium having a characteristic brush border.

Mandibles In an insect mouth, these are a pair of unjointed, hard, biting jaws which articulate with sides of the head capsule. On their inner edges, they bear prominent teeth like serrations which are opposed to each other and thus the mandibles are concerned with the cutting and crushing of food into small bits.

Mandibular sclerite *See* optic tubercles of housefly.

Mandibulate mouth parts Cockroach has this type of mouthparts, that is, they are adapted for biting and chewing the food.

Mange Scabies disease. *See* Animal Science.

Mango hoppers A serious insect pest of Mango–*Idioceros* at kinsomi and *I. clypelis*. For control, sprays of crude oil emulsion, kerosene emulsion, fish oil soap, fish-oil-rosin-soap and tobacco decoction are helpful. Small, greyish brown, wedge-shaped insects that suck along with their nymphs from new shoots, buds and flowers, causing the inflorescence and tiny fruits to dry up; they secrete a sticky substance on which a sooty mould develops. Avoid waterlogged and damp conditions in mango orchards; spray with 0.25 per cent DDT early in spring when infestation starts.

Mango mealy bug of jack fruit (*Artocarpus integrifolius* Auth.) Same as for mango.

Mango mealy bug of mango (*Mangifera indica* L.) *Drosicha mangiferae* Green. A soft, white, slow moving insect, seen clustering round young shoots and inflorescence from December to April from which it sucks juice; shoots and flowers dry up; it attacks a number of other fruit trees. Apply sticking bands round trunk of trees in cold weather; spray with 0.06 per cent malathion or 0.05 per cent parathion. Spray @ 2–5 gallons per grafted tree and 5–10 gallons per seedling tree in early stages of infestation.

Mango shoot gall maker of mango (*Mangifera indica* L.) *Apsylla cistellata* Buckton. Round, pale yellow nymphs of a small, winged, sucking insect; produce cone-shaped galls in mango shoots; shoots damaged and tree remains stunted. Collect and burn the galls.

Mango stem borer of fig (*Ficus carica* L.) Same as for mango.

Mango stem borer of custard apple (*Annona squamosa* L.) *Olencecambtus biologus* Fab. A beetle whose grub bores into the stem and kills affected branch. Remove and burn infested branches; fumigate holes with petrol or carbon disulphide.

Mango stem borer of mango (*Mangifera indica* L.) *Batocera rufomaculata* De Geer or B. *rugra linn.* A grub of a large, long-homed, brownish beetle that bores into and makes tunnel in branches and stem; the beetle feeds on bark of twigs and lays eggs in the bark. Fumigate holes and tunnels with petrol or ED/CT mixture.

Man's insect parasites These species infest man. (1) the head louse (*Pediculus humanus capitis*) lives on the hair of head. (2) the body or clothes louse (*Odiculus humanus corporis*) is dangerous as well as disagreeable since it transmit typhus fever, relapsing fever and trench fever. The crab-louse (*Phthirius pubis*) lives in the skin and the hair of the pubic and inguinal regions. The bed bug and the fleas are other enemies of man. They also attack domestic animals.

***Marasmia trapezalis* Guen** *See* Leaf roller of Maize.

***Margaronia caesalis* W** *See* Shoot borer of Jack Fruit.

Mascular system It is highly developed in Arthropoda with numerous separate muscles composed of finely striated quickly contracting fibres.

Maxillae These are situated just behind and below the mandibles. Each maxilla consists of a blast part of two sclerites, the cardo and stipes bent at an angle to each other. From the outer border of the stipes arises a five jointed maxillary palp at the base of which is a small sclerite known as palfiper.

Maxillary palp *See* malillae.

Meal snot moth, *Pyralis farinalis* Linn. A light brown moth with fore wings dusky white in middle and chocolate brown at edges; caterpillars live in silken tubes constructed from foodstuffs they infest; attacks coriander, cumin, sago, mustard seed, potato, oatmeal, bran, etc.

Mealy bug of citrus *Seudococcus citri.* Can be controlled by emulsion. The giant mealy bug of Mango is white; bug is Drosicha (Mnophlebus) stebbingi.

Mealy bug of cacao (*Theobroma cacao* L.) *Pseudococcus citri* R. Pods are infested with soft scale insects. Same as for scale insects on tea.

Mealy bug of custard apple *Pseudococcus lilacinus* C.

Mealy bug of custard apple (*Annona squamosa* L.) *P. lilacinus* C.

Mealy bug of custard apple (*Annona squamosa* L.) *Pseudococcus virgatus* C. Colonies of small, reddish white mealy bugs infest leaves, shoots, buds and fruits, sucking sap from them. Spray with 0.03 per cent parathion @ 2–4 gallons per tree.

Mealy bug of pineapple (*Ananas sativas* Schult.) *Pseudococcus bromeliae* B. Colonies of reddish white mealy bugs infest shoots and fruits, and suck sap from them. Spray with 0.025 per cent parathion @ 60–80 gallons per acre.

Phenacoccus inceryoides Gr. They infest shoots and stalk. Spray with 0.03 per cent parathion @ 2–5 gallons per tree.

Mealy bug of sugarcane (*Saccharum officinarum* L.) *Saccharicoccus sacchari* (Ckll.). A small pinkish scale insect that suck juice from stem at lower nodes. Destroy infested parts; spray with 0.25 per cent parathion.

Mealy bug of tomato (*Lycopersicon esculentum* Mill) *Pseudococcus virgatus* F. A scale insect that covers up the whole plant and sucks the juice. Remove infested shoots; if infestation is severe, spray with 0.06 per cent malathion @ 60–80 gallons per acre.

Mealy scale of guava (*Psidium guava* L.) *Pulvinaria psidii* Maskell. Colonies of mealy bugs cover the leaves and suck juice causing a sooty mould to develop. Spray with 0.03 per cent parathion or diazinon @ 1–2 gallons per tree.

Mealy wings of sugarcane (*Saccharum officinarum* L.) *Aleurolobus barodensis* M. Also known as sugarcane white fly; small sucking insect of pale green colour; found on leaves; both adults and nymphs cause damage. Remove attacked leaves promptly; spray with 0.25 per cent BHC or 0.05–1 per cent parathion @ 100–150 gallons per acre.

Mechanism of breathing in insects Oxygen diffuses inwards and carbon dioxide outside through the cuticular covering of the whole body. This is possible for the chitinous exoskeleton, which is impervious to water, is previous to gases.

Mecoptra It includes insects having complete metamorphosis, chewing mouth parts and long and slender antennae. The head is prolonged into a beak and there are two pairs of similar membranous wings which the veins divide into equal areas, *e.g.*, scorpion flies.

Mediproboscis *See* mouth parts of housefly.

***Melolontha nepalensis* Bl.** *See* leaf and fruit eating beetles.

Mentum The middle segment of the maxillae of insects.

Mesentric *See* mesentron.

Mesentron The middle portion of the alimentary canal is the mesentron or mild-gut. It is a tubular structure of approximately uniform diameter. From its anterior end arise 7–8 blindly tubular mesentric or gastric caeca. The slightly coiled alimentary canal of a insect, consist of (1) a long stomodeum, (2) a short mesentron, and (3) proctodeum.

Mesolegs *See* legs of cockroach.

Mesothorax *See* thorax.

Metamorphosis In Arthropods, fertilisation is internal and most of them undergo a more or less extensive transformation *i.e.* from the egg to adult (egg-larva-adult). The egg hatches a larvae which develops by definite stages into the adult forms; the metamorphosis being accompanied by frequent moulting the inelastic chitinous exoskeleton.

Metamorphosis in cockroach The embryonic development is completed within the ootheca. After some time the eggs enclosed in the ootheca hatch and from these emerge nymphs which are much like the adults, except in size, colour and absence of wings. During its further development and nymph moults ten times as a result of which by gradual development it reaches the adult stage.

Metamorphosis of housefly It is a series of changes through which the insect passes in its growth from the egg through the larva and the pupa to adult. The pupa remains immobile and apparently inactive, yet within it is a science of tremendous change in development activity. Most of the larval organs with the exception of the central nervous system are broken down largely by the action of the phagocytic blood corpuscles into an elemental cream. The process is called histolysis. The elemental cream forms food for the nourishment of certain groups of cells which were clearly set aside and these groups of cells are called imaginal discs or buds. These are distributed throughout the larval body in situations appropriate for the development of the adult organs. During favourable conditions, the entire reconstruction process histogenesis may be completed within four or five days but during cold weather, it may take several months. At the end of completion of metamorphosis, the imago of the fly has to free itself by means of a bladder-like structure, the ptilinum, which is situated on the front part of the head.

Metamorphosis of mosquito The scrapping of the larval organs is generally (Histolysis) affected by the phagocytes. But while this breakdown of the larval organs is going on the rebuilding process or histogenesis is also carried out. After a period of one to three days, a pupa of mosquito is transformed into a perfect adult or imago. When the change is about to take place, the pupa becomes more buoyant, rises to the surface of water and lies parallel to the surface. After a minute or two, the pupal skin splits along the back of thorax and the mosquito emerges and projects above the surface of water. The insect's body is now carefully withdrawn from the pupal case, which is used as a float.

Metamorphosis of a silkworm moth The moth lays its eggs in hundreds on the mulberry leaves, and to protect the eggs from falling off, the moth covers them with a gelatinous material by means of which the eggs stick to the leaves. At the time of hatching out, the eggs turn brown from white. The silkworm which is hatched is called caterpillar or larva. It starts at once to feed voraciously on the mulberry leaves. After about four days, the silkworm stops eating and appears listless. Eventually its skin bursts and it emerges with a new skin. Then it begins eating and growing again. After about a week, the worm repeats this process and does this four times. Also *See* silk glands.

Metamorphosis in butterfly After mating, the female lays eggs in batches of from 6 to 100 on the underside leaves of the food plants of the larva. The eggs are attached by a sticky secretion. It contains reserve food material in the form of yolk and a small hole or micropyle on the top. This mycropyle allows the sperm to pass and when the eggs has been fertilized the same allows air to reach the developing embryo. After about 6 to 8 days, young larave called caterpillars, eat their way out of the eggs' shells and eat their remains. *See* caterpillars.

Metathorax *See* thorax.

Method of combating fly pest (1) Prevent the breeding of flies. (2) Kill the adults as far as possible. (3) Prevent the transfer of infection by flies.

Method of combating housefly pest (1) Swat and kill every fly. (2) prevent their breeding. (3) prevent the transference of infection by flies.

***Metriona circumdata* H.** *See* Leaf beetles of Sweet Potato.

***Microtermes obesi* (Anandi) H.** *See* Termites of Wheat.

***Microtermes obesi* Himgr.** *See* Termites of Sugarcane.

***Microtermes* sp.** *See* Termites of Tea.

Mites Minute animals with four pairs of legs that attack grain and grain products with high moisture content; they cannot fly, and the sources of infestation are the stored produce, storage receptacles and storage place already infested by them.

Mites of animals They produce a specific contagious disease known as mange (or scabies, scab or itch). These small insect parasites constitute a very large group. Each species of domesticated animals has its own peculiar species of mange mites. Mites are responsible for the condition known as mange (scabies) in cattle, sheep, horses and swine. The disease spreads most rapidly among young and poorly nourished animal. Mites can be controlled by spraying or dipping infested animals or infested premises with suitable insecticides. A spray of dip of Lindane 0.05–0.06 per cent dip will help.

Mites of litchi, *Eriophyes* spp. Causes leaf curl or ermose. A serious pest. Sometimes the whole crop or nursery may be destroyed. Solutions of Nicotine sulphate and whale-oil soap gives good control of the mites.

Mites of mango (*Mangifera indica* L.) *Tyrophagus asiaticus* Evans, *Eriophyes* sp. Mites attack mango shoots and inflorescence, causing malformation and giving a bunchy appearance. Remove all infested inflorescence and shoots up to one foot below the point of attachment to the host.

Mole cricket *Grytalpa africans* damage many crops.

***Monolepta signata* O.** *See* Leaf eaters of Colocasia.

Mosaic image Same as apposition image.

Mosquito It is a member of a very large group of Insecta known as Diptra to which also belong the familiar housefly. It belongs to family Culicideae. The common species are *Culex pipens* and *Anopheles* sp.

Mosquito sound The familiar buzzing sound of mosquito is produced in two ways. The deeper note is due to rapidity of the wing strokes of which there are many hundreds per minute. The shriller note which is apparently confined to the female is due to the vibrations of the two tense membranes at the openings of some of the anterior spiracles or breathing pores.

Moth Characteristics of moth are (1) antennae are tapering and often feathery (2) the body is often stout and hairy (3) resting position of wings horizontal; fore wings more or less covering hind wings (4) usually fly at dusk and after (5) eggs flat or rounded (6) pupa usually smooth, often growth.

Moulting *See* larva of housefly.

Mouthparts Joint appendages on the head of arthropods, modified in various ways for dealing with food. They consist of the labrum (upper lip), which is a single plate; a pair of mandibles (upper jaws), which have serrated edges; a pair of maxillae (lower jaws), which also have serrated regions; and the labium (lower lip), which is a single plate. The labium and maxillae have sensory feelers (palps) concerned with the tasting of food. The basic system of mouthparts found in the more primitive groups (*e.g.* cockroach and locusts) is highly modified than in other groups to suit the mouthparts for one particular kind of food. For example, the tubular mouthparts of butterflies and moths are adapted for sucking nectar; the piercing and sucking mouthparts of mosquitoes are adapted to feed on blood or plant juices. The labium is equivalent to the second maxilla of crustaceans in which it may also be used for producing respiratory as water currents. *See* also chelicera, pedipalps.

Mouthparts of butterflies The proboscis which normally lies coiled beneath the head consists of two processes, each bearing teeth-like processes on the inner side. These teeth help to look the two halves of the proboscis together forming a tube along which nectar from flowers can be sucked. On the head of the caterpillar, there are the powerful biting jaws which move transversely and help them in feeding on leaves of plants.

Mouthparts of cockroach The mouth parts of cockroach are labrum, a pair of mandibles, a pair of maxillae, a labium, a hypopharynx.

Mouthparts of honeybee These are modified so as to enable the bees to obtain nectar from the bases of flowers. Typical mandibles are present in addition to a long sucking proboscis. The very long maxillae together provide a protecting sheath for an equally long sucking tube which is formed by the fused inner parts of the labium and is known as ligula. The end of the tube is open and its tip is slightly expanded ventrally to form a minute spoon-like bowl. When not in use, the whole structure is packed away in a groove below the head. From the tongue, the nectar passes into the crop or honey sac, where it is mixed with the secretion from the salivary glands and undergoes chemical changes before being regurgitated as honey and stored into hive.

Mouthparts of housefly In has sponging type of mouth parts on the lower side of head; is fleshy and retractile proboscis, which is formed by the labium. The latter consists of three parts (*i*) the large basal part, the basiproboscis or rostrum, (*ii*) the stout cylindrical mediproboscis or haustellum, and (*iii*) the distal segment or distiproboscis. The hausrellum is distally attached to the (movable) apex of the rostrum and can be folded back below the rostrum and hidden beneath the head. The distiprobosis is expanded into a pair of oval membranous sponge-like organs, the labellae, which are traversed by a series of furrows or channels which narrowly open all along the exposed end. These fine channels are the pseudotrachae. They converge at one point on the labella and from the point the liquid food is drawn up through the food channel into the oesophagus. The food channel is formed by the labrum-epipharynx-hypopharynx. These are the only two styles that are present and incapable of piercing the skin. The mandibles are wanting and the maxillae are represented by only a pair of unjointed maxillary palps, which lie in front of the proboscis.

Mouthparts of mosquito These form a very efficient apparatus for piercing and sucking. There are only three appendages *viz.* a stout middle process with another on either side of it. The side processes are called palps. In the male, the palps are as long as the proboscis but in the female the palps are much shorter. The proboscis is best developed in the female and consists of six lancet like processes, lying very close together in a grooved sheath. The proboscis of a female mosquito consists of (*i*) a pair of mandibles, which are delicate needle like processes sending in tiny blades. (*ii*) a pair of maxillae (1st mamilla) which are also needle like but with delicate saw like at their tips.

***Musca* spp. (including house flies)** *Musca domestica,* the housefly of the hills in India; *M. nebulo,* the bazaar fly, and M. *vicina,* the common housefly of the Indian plains are non blood-sucking, but they are known to carry many species of disease germs on their legs and other parts of the body. Several species of Musca are haematophagous, as described earlier, while others are true biting flies. A notable example of the latter is *M. crassirostris,* which attacks cattle with pertinacity and causes considerable annoyance to them. Its mouthparts are not adapted for puncturing the skin, but for tearing the skin to the extent necessary to enable it to obtain a meal of blood.

Muscideae House fly and stable fly. *See* Animal Science.

Mustard aphid of crucifers and mustard (*Brassica campestris* L.) *Lipaphis* (Rhopalosiphum) *erysini* Davis. A greenish white aphid; affected leaves get curled, and plants wither and die. Clean cultivation; spray with 0.05 per cent BHC or 0.05 per cent malathion @ 80–100 gallons per acre.

Mustard saw-fly of mustard (*Brassica campestris* L.) *Athalia proxima* Klug. A blackish caterpillar that attacks all crucifers; feeds on leaves of young seedlings;

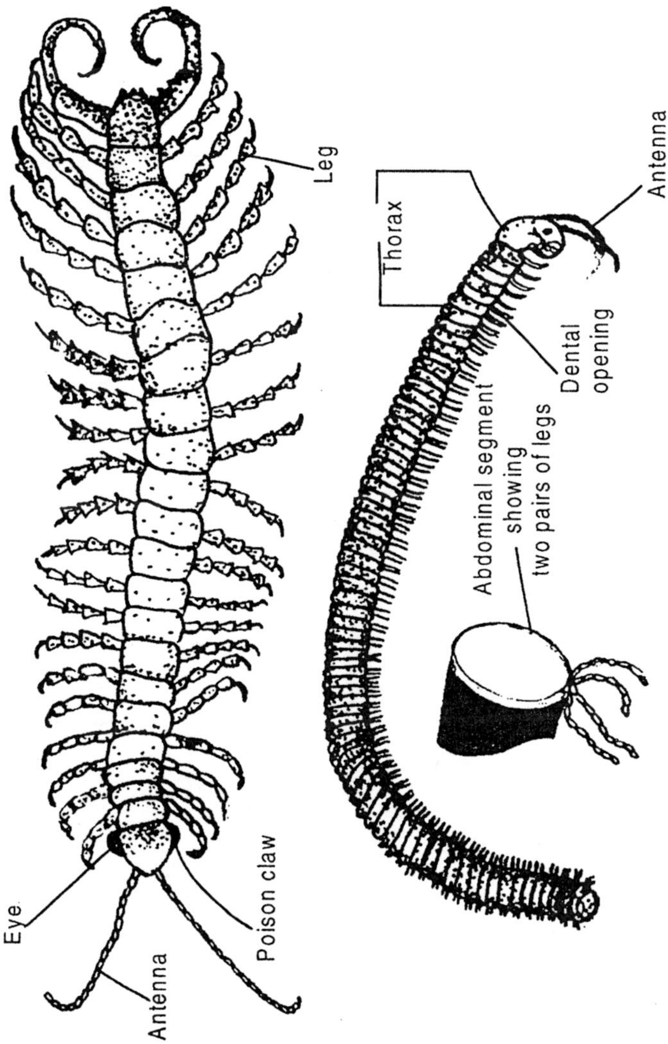

Myriapoda
(A) Millipeda, (B) Centipede

curls and falls on the ground when touched; the adult is a black fly that lays eggs singly inside the leaf tissues. Remove larvae by handpicking if infestation is light; dust with 3 per cent BHC @ 15-20 lb. per acre or spray 0.1 per cent DDT or BHC @ 80–100 gallons per acre.

***Myllocerus* sp.** *See* grey weevil of soybean.

***Myllocerus* spp.** *See* grey leaf weevil of cotton.

***Myzus persicae* Sulz** *See* aphid of linseed and potato.

Myriapoda It includes elongated anthropoda with a well defined head region, trachaeal tubes and many similar leg-bearing segments. There is one pair of antennae. The group is divided into two main per segment and those having one pair per segment (Chilopoda).

N

Neck of insect *See* head of insect.

Nectar *See* mouthparts of honeybee.

Needle like processes *See* mouthparts of mosquito.

Neotermes **sp.** *See* termites of mango.

Nervures In insects, each wing is strengthened by a framework of branching veins called nervures.

Neuro-sensory system in insects In cockroach, the central nervous system being of the usual annulate type, resembles in its main features, that of an earthworm.

Neuroptra This order includes insects which have complete metamorphosis chewing mouthparts, four similar membranous wings, usually with many veins and cross veins. Larvae are carnivorous, some with suctorial mouthparts. Tracheal gills usually present on aquatic larvae, *e.g.*, lacewing and alder flies.

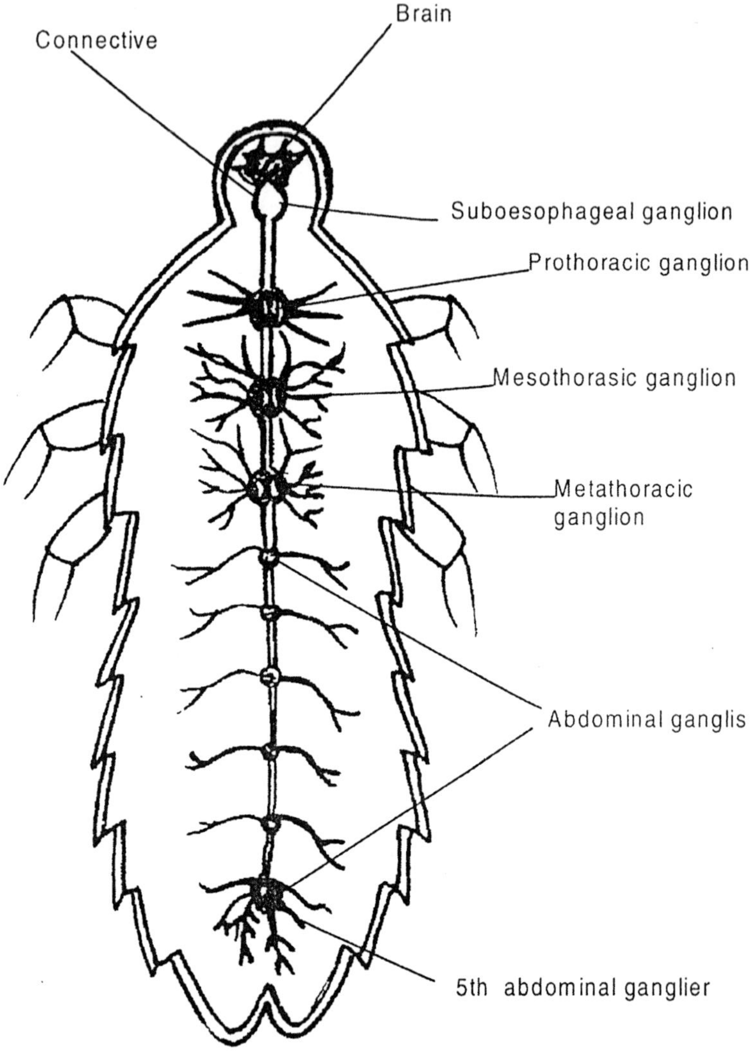

Nervous System of Cockroach

Nervous system In *Arthropoda,* it is annulate type, consisting of paired dorsal ganglia over the mouth, a ring or collar round the gullet and a ventral chain of ganglia.

Nicotine insecticides In this class of insecticides, the most important material is tobacco, the effective agent being the alkaloid nicotine. It acts as a repellent as well as a contact insecticide. Tobacco decoction, nicotine sulphate are some insecticide in common use.

Notum *See* Tergum.

Nuptial flight *See* flight to death.

Nut weevil of mango (*Mangifera indica* L.) *Cryptorhychus mangiferae* Fabr. C. gravis Fabr. The grub bores into the stone, feeds inside and comes out through the pulp damaging the same. Spray with 0.2 per cent DDT twice or thrice at 15-20 days interval after the formation of the young fruits.

Nymphs and adult females suck the sap *See Aspidotus cydoniae* Comst.

Nymph *See* instar.

O

Ocellar spot *See* Fenestra.

Ocellus A simple eye as in an insect, object designating pupae (of Lepidoptera) in which the wings and legs are fused to the body. Compare coarctate, exarate.

A simple eye of some invertebrates (*e.g.* arthropods). It consists of a group of light sensitive cells (photoreceptors). Ocelli are concerned with perceiving the direction and intensity of light and do not form an image.

***Ochyromera artocarpi* M.** *See* bud weevil of Jack Fruit.

***Odoiporus longicollis* Ol** *See* Stem borer of Banana.

Odonota In this order of Insect, are included slender insects often with large wings which are similar in size and venetion. Have large eyes and aquatic nymphal stages. Examples–dragon flies. The order or insects that contains the dragon flies, brightly coloured carnivorous insects with a long thin abdomen and two pairs of equalised elongated wings. The head bears large compound eyes, reduced antennae, and biting mouthparts and the legs are set well forward for catching the smaller insects that they prey on. The acquatic nymphs are also carnivorous, with an enlarged labum (the mask) for catching prey.

***Odontotermes obesus* R.** *See* Termites of Sugarcane.

***O. assumathi* Hlmgr** *See* Termites of Sugarcane.

***O. taprobanes* Wlk** *See* Termites of Sugarcane.

***Oecophylla amargadian* Fabr.** *See* Red ant of Mango.

Oesophagus In cockroach, this passes through the neck into the thorax where it dilates to form a large thin walled and pear-shaped crop, which extends through the thorax far into the abdomen and is by far the largest part of the alimentary canal.

Oestridae Warble or nasal fly. *See* Animal Science.

***Olenecambtus biologus* Fab** *See* Mango stem borer of Custard Apple.

***Oligonnychus indicus* Hirst** *See* mite of Sugarcane.

Ommatidium A long narrow pyramidal body in an insect eye, which is in itself a complete organ of vision. Owing to its minute size and structure, it has very limited range and needs, therefore to be supplemented by other similar visual elements surrounding it. Each such visual element is the ommatidium and very large number of ommatidia over two thousand-placed side by side, form the entire eye.

Omnivorous insects Insects like cockroach are omnivorous creatures and can eat any kind of organic matter *viz.*

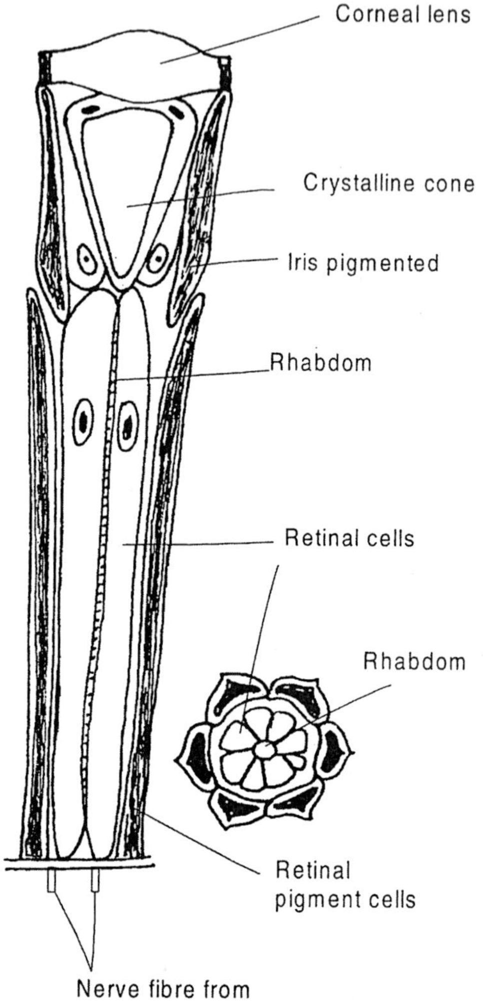

Longitudinal section of a single ommatidium together with the cross section through the rhabdom

bread, meat, paper, cloth, wool, horn, dead bodies of its own fellows and even its own cast away exoskeleton.

Oncocephata tugerculata **O** *See* Leaf beetles of Sweet Potato.

Onion thrips of onion and garlic (*Allium cepa* L.; *A. sativum* L.) *Thrips tagaci* L. Small sucking insects with fringed wings that feed on leaves and lacerate the tissue. Spray with 0.06 per cent Nicotine sulphate of 0.1 per cent DDT @ 60–80 gallons per acre.

Ootheca *See* hypogynium of cockroach.

Oothecal chamber See female gonopore, the hinder part of the boat shaped cavity. In a female cockroach, the 8th and 9th sterna of the abdomen get pushed in to form the inner wall of the oothecal chamber.

Ophiderus materna passes the larval stage on *Tinospora cordifolida* The elimination of these hosts and possibly of some others is the most effective control measure.

Optic nerves These nerves arise from the large rounded parts of the cerebral ganglia arise the optic nerves while the antennary nerves arise from the lower portions of these ganglia.

Optic tubercles Each oral lobe in housefly larva bears two conical sensory optic tubercles and its ventral surface is traversed by a number of food channels. The mouth opens on the ventral surface between the two oral lobes and its anterior border is occupied by a black hook-like mandibular-sclerite which helps only in locomotion and not in mastication only liquid food is absorbed.

Oral lobes The housefly larva has oral lobes which are united posteriorly on the dordal side.

Orange beetle borer of citrus crops *Stromatium barbatum* Fabr. The grub bores into branches; the beetle lays eggs in summer and the grub continues to feed inside its tunnel till next spring. Cut and burn affected branches; fumigate borer holes with petrol or ED/CT mixture.

Orange hair streak An insect pest of oranges, *Tarcus theophrastus*.

Orange leaf caterpillar *Tonica ziziphi* is a leaf eating caterpillar.

Organge spiny fly *Aleurocanthus spiniferus.* To control, rosin soap in winter and rosin compound in September and also tobacco decoction are effective.

Organs of taste in insects These are present on the tips of the maxillary and labial palps. Auditory sensillae are believed to be present on the anal cerci probably to help detect mechanical strains in the chitin of the exoskeleton.

Orthoptera It is a large group of mainly tropical insects such as grasshoppers, locusts and crickets.

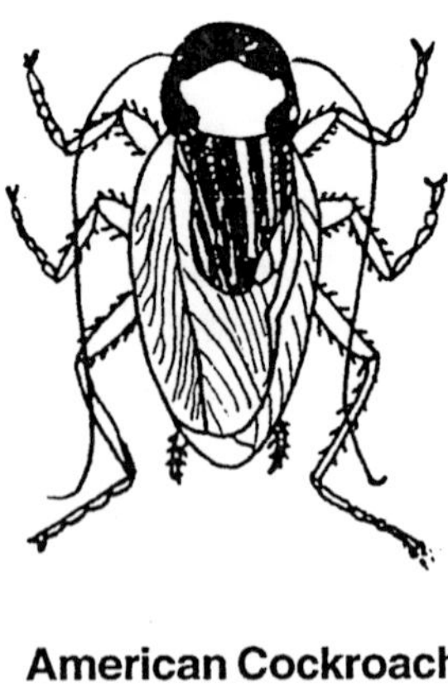

American Cockroach (Orthoptera)

Walking Stick (Orthoptera)

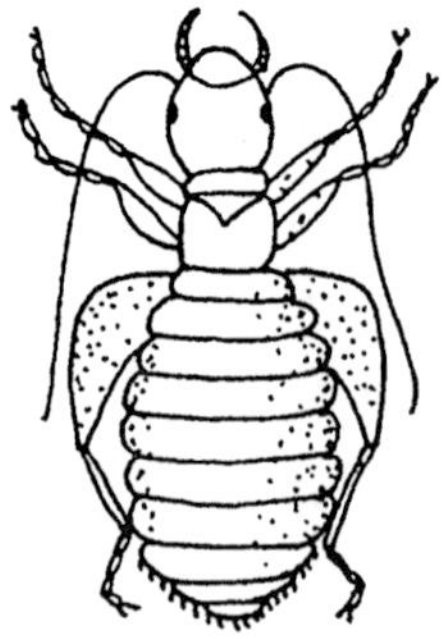

Book Louse (Corrodentia)

Oryctes rhinocerus **L.** *See* Rhinoceros beetle of Coconut.

Oryzaephilus surinamensis **Linn** *See* Saw-toothed grain beetle.

Osin compound Also called rosin compound. A fish oil soap containing rosin is also available, and the rosin in it makes it more adhesive. When soap is used alone, about ½ kg to three gallons of water should be used. A compound made by adding two pounds of powered rosin to one pound of washing soda in one gallon of water, and again boiling, is sometimes used.

Ostia *See* pericardial sinus.

Ostium A mouth-like opening; for example, any of the lateral openings in the heart of an arthropod or any of the openings through which water enters the body of a sponge.

Otheis (Ophederis) fullonica Moths which fly at night and suck the juice from fruits occur all over the Citrus districts of Asia, and cause very serious damage. The moths pierce the rinds of ripening fruits, thus making possible the entrance of decay organism. Ordinarily the fruit falls off within a few days. The most common species is this one with grey upper and bright orange and black lower wings. The larvae feed on the weed *Cocculus hirsutus*.

Other species of insects Certain species of ants are very troublesome, for once they start an attack, they will persist in it regardless of loos of life of their own kind. Keeping the hive stand legs in water is the preventive. Dragon flies, praying mantids, as also spiders, eat straggler bees when they can find an opportunity to do so.

Othreis fullonia **El.** *See* Fruit sucking moth of Citrus Crops.

Ovaries The female cockroach has two ovaries, one on either side of the median line lying embedded in the fat body in the 4th to 6th abdominal segment. Each ovary consists of long beaded tubes known as ovarioles or egg tubes.

Oviduct The two short and wide but muscular oviducts unite to form a common oviduct which opens into the vagina by a vertical slit, the female gynopore.

Oviparity The condition that involves laying of eggs.

Ovipositor An egg-laying structure at the hind and of the abdomen of female insects, formed from modified

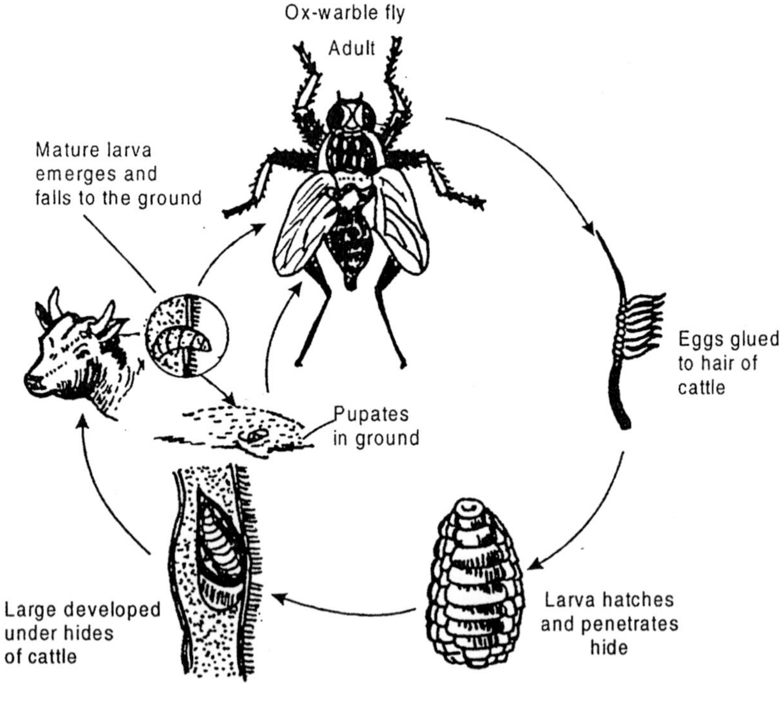

Life Cycle of Ox-warble Fly

P

Pachydiplosis oryzae **W.** *See* Paddy gall fly of rice.

Paddy bug of rice (*Oryza sativa* L.) *Leptocorisa varicornis* Fabr. Bugs suck the milky sap of tender grains; sometimes crop destroyed completely. Clean cultivation burn trash in fields or shake plants to drop young nymphs in water; dust with 5 per cent BHC at 12 to 15 lb. per acre.

Paddy cutworm of rice (*Oryza sativa* L.) *Cirphis unipunctata* Haw. Caterpillars appear after heavy rains or early floods. Trap caterpillars in grass heaps or plough up infested fields; dust with 5 per cent BHC at 15 to 20 lb. per acre.

Paddy gall fly of rice (*Oryza sativa* L.) *Pachydiplosis oryzae* W. Maggots attack base of growing points and produce silvery gall (silver shoot). Remove grasses from bunds and use light traps for adult insects. Spray with 0.25 per cent DDT or 0.02 per cent endrin @ 60–80 gallons per acre.

paired appendages. It is frequently long and needle-like to enable the piercing of animal and plant tissues to lay eggs. In wasps, bees, and ants, it is modified into a sting.

Ovipositor of bees Both queens and workers of honey bees posses a sharp tube which can be protruded from tip of the abdomen. In queens, this tube is used for the deposition of the eggs and hence it is called ovipositor.

Ox-warble fly Cuts holes in the skin of cattle in order to obtain air and thus their hides are badly damaged.

Oxya velox fbr. (Leptispa Pygmaea b) *See* paddy grasshopper of rice.

Paddy grasshopper of rice (*Oryza sativa* L.) *(Oxya bvelox fb.), Leptispa pygmaea* B. and *Paddy jassid (Nephotettix bipunctatus* Fb.). Minor pests. Use light traps of jassides; dust with 5 per cent BHC at 15–20 lb. per acre in all cases.

Painted bug of crucifers *Bagrada cruciferarum* Kirk. A black and orange bug that sucks the sap; eggs are laid on leaves or in loose soil; newly hatched nymphs have bright orange colour. Clean cultivation; spray with fish-oil rosin soap at 30 gallons per acre to kill nymphs; dust with 5 per cent BHC in case of severe infestation @ 15 to 20 lb. per acre.

Palm weevil *See* borer.

Palpizer This is a small sclerite situated at the base of the maxillary palp of the insect mouthpart.

Palps *See* mouthparts of mosquito.

Palvillae These are the well developed glandular pads between the claws of the legs of the housefly. These bear innumerable hollow hairs through which sticky substance exudes. It is this sticky substance that enables the fly to walk lupside down and also to climb smooth surfaces.

***Papilio demoleus* Linn** *See* lemon butterfly of Citrus Crops.

Paraglossa The outer portion of the insect labial pulp. *See* Prementum.

Parapprocts *See* anal cerci of cockroach.

Para rubber termite of para rubber (*Hevea brasiliensis* Muell.) *Ghyptotermes dilatatus* Havil. Termines attack the woody part of the plants. Mixing aldrin 5 per cent dust in the soil around the trees.

Parasite organisms These are the natural enemies of insects such as predators *Cicustella* spp., and *Canthiconidia furcillata* spp.; are also recorded as natural enemies of the pests. Also *See* Plant Science.

Parasitic insects *See* Animal Science and Crop Science.

Par green A stomach insecticide. It is the aceto-arsenite of copper and contains about 45 per cent of water soluble arsenic which is the active ingredient of this compound.

Peach scale of peach (*Prunus persica* Batsch) *Pseudaulacapis pentagona* Targ. Sucks juice from twigs to which they remain attached. Spray with fish-oil rosin soap (2 lb. in a gallon of water) 0.03 per cent parathion @ 1.2 gallons per tree.

Peas semilooper of gram (*Cicer arietinum* L.) *Plusia orichalcea* Fabr. A greenish semi-looper that feeds on leaves. Caterpillars may be handpicked or sprayed or dusted as for gram caterpillar.

Pedicel In insects, it is second segment of the antennae.

Pediculus humanus capitis This the head louse of man and lives of head.

Pediculus humanus corporis This is the body or clothes louse of man, is dangerous as well as disagreeable as it transmits typhus fever, relaping fever, and trench fever.

Pedipalps A pair of appendages on the fourth segment of the head of arachnids. They function in the place of antennae (absent in this group) in spiders and are used for grasping prey in scorpions. The segment nearest to the body bears a hard plate used for chewing food.

Pempheres offinis **F.** *See* stem weevil of Cotton. Pentastomida *See* Animal Science.

Pericardial cinus Also called cavity. The heart is made of 13 segmentally armaged funnel shaped chambers, each of which is placed in communication with the pericardial sinus or cavity by a pair of laterally placed ostia which permits the entry of colourless blood known as hymolymph.

Pericardial haemocoel The thorax and abdomen in cockroach are divided by two delicate membranous horizontal diaphragms into three regions: (*i*) pericardial haemocoel, (*ii*) perivisceral heomocoel, and (*iii*) sternal haemocoel. The heat lies in the dorsally situated pericardial haemocoel, which is separated from the previsceral haemocoel by the diaphragm or pericardium. To its floor are attached intersegmental alary muscles.

Pericardial sinus *See* circulatory system in insects.

Perigoea copensis **G.** Caterpillar of Safflower.

Periodic parasites Same as visiting parasites.

Peripheral nervous system From each ganglionic mass arise nerves which pass to the various organs of the segments in which they lie and form the peripheral nervous system.

Periplenata americana It is a native of tropical America and is common in India. This cockroach is big in size. *See* Cockroaches.

Peritrophic membrane The food content of the mesentron of cockroach is separated from its epithelium by a thin membrane known as peritrophic membrane.

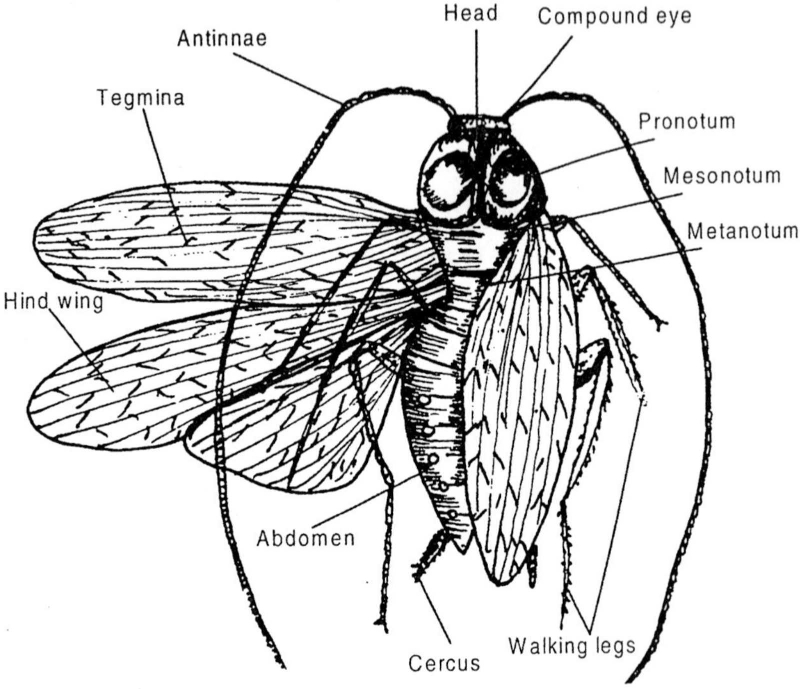

Dosal View of *Periplaneta americana*

Phallic The gland which is a large elongated sac-like structure lies beneath the urticular gland and ejaculatory duct and opens by the side of the male gonopore in insects.

Phallomere *See* genetalia in insects. In cockroach, they are the parts of the ninth segment.

Pharynx In cockroach, the mouth which is situated at the base of the preoral cavity lies the tongue-like hypopharynx, leading of the tubular pharynx which passes vertically upward through the nerve-collar into the oesophagus.

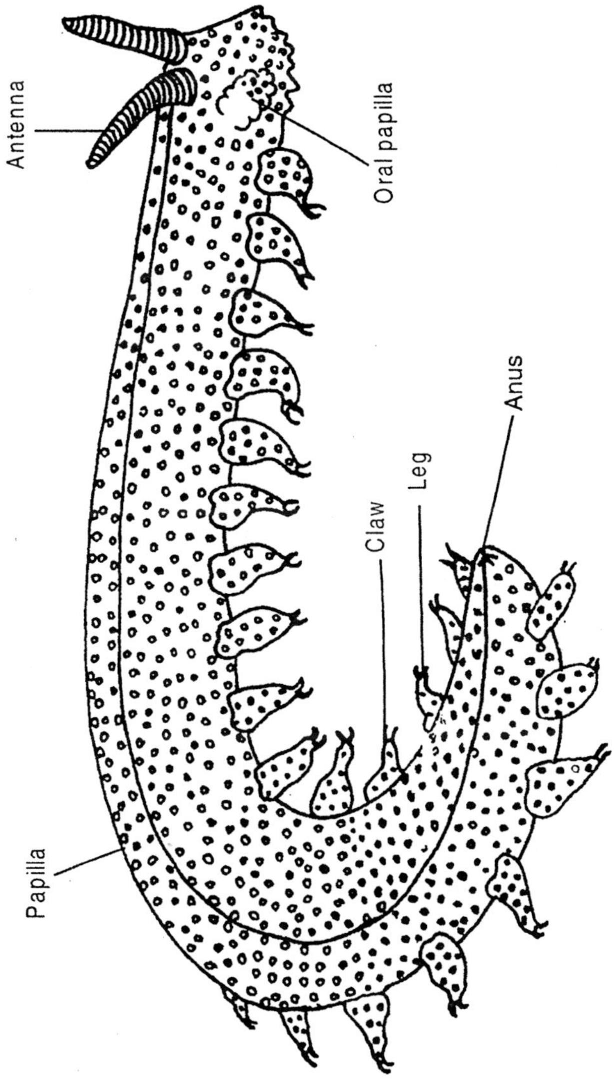

Peripatus

Phaseoli gyll *See* Pulse beetles.

Phases of locusts Unlike grasshoppers, locusts exist in two phases–(1) solitary, and (2) gregarious. In the solitary phase, they are inactive, live and bread as scattered individuals, and are relatively harmless. In the gregarious phase, they are very active, acquire the habit to congregate and form swarms, breed rapidly and in adult stage, fly long distances in migration even across seas. Climate, weather, soil, vegetation and their mass destruction or gained by man interfere with their habit and capacity to undertake migratory fights.

Phasmida This order of insecta consists of insects which are often long with or without wings such as stick insects.

Phenacoccus insolitus **G** *See* Sucking insects of Brinjal.

Phenacoccus iceryoides **G** See Mealy bugs of Sapota.

Phthirius pubis It lives in the skin and the hair of pubic ad inguinal regions.

Phychodicae Sand flies. *See* Animal Science.

Phyctia clientalla **Z** *See* Leaf caterpillars of Brinjal.

Phyllexora **spp.** Grapes are flourished in France and other warmer parts of Europe for centuries before the discovery of America and the subsequent discovery of American species of grapes. When the American grapes were taken to France, it carried with it the insect Phylloxera. This insect soon began to attack the European grapes and caused great damage. For a time it seemed as if the grapes industry might be wiped out of Europe. But the American grapes through long association with this insect European grapes on the

American roots, it was found that they could be grown without danger. The somewhat more costly form of propagation was much more than justified.

Phyllocnistis citrella **Stainton** *See* leaf miner of Citrus Crops.

Phyllotreta cruciferae **Goeze** A small dark grey flea beetle that damages leaves by making holes. Dust with 5 per cent BHC @ 15 to 20 lb. per acre.

Piercing stylets See mouthparts of mosquito.

Piercing and sucking mouthparts These are found in mosquitoes which are adapted to feed on blood and plant juices.

Pink bollworm of cotton (*Gossypium* spp.) *Platyedra gossypiella* Saund. A pale or deep pink caterpillar that bores into bolls and seeds and pupates therein; the adult is a small blackish brown moth. Treat seeds before sowing by sun-heat treatment; remove and destroy fallen buds and bolls. Spray as under spotted bollworms.

Pink borer of maize (*Zea mays* L.) *Sesamia inferens* Wlk. Phadka grasshopper. *See* under wheat.

Pink stem borer *Sesamia inferens* This lepidoptrous insect damages wheat and maize crops and several other crops also. Handpicking of infested fields and their destruction reduces the incidence for next crop.

Plantulae *See* legs of cockroach.

Platyedra gossypiella **Saund.** *See* Pink bollworm of Cotton.

Plecoptra This order of class insects, have two pairs of large similar wings and are found near water, *e.g.*, stoneflies.

Pleuron In cockroach, the central sternum is joined at the sides by a soft cuticle known as pleuron. *See* thorax.

Plodia interpunctella **Hubn** *See* Indian meal moth.

Plusia orichalcea **Fabr.** *See* Peas semi looper of Gram.

Plutella maculipennis **C.** *See* Diamond black moth of Crucifers.

Pluteus A form of dipleurula larva characteristic of brittle stars and sea urchins, in which the ciliated band is continuous, with a small pre-oral lobe and well developed post anal lobes supported by calcareous ribs. *See* dipleurula.

Podical plates In insects, the anus is situated between two podical plates or paraprocts. *See* anal cerci of cockroach.

Poison Baiting of locusts. *See* locust control measures.

Pollen basket *See* legs of honeybee.

Pollen brush *See* legs of honeybee.

Pomegranate butterfly A serious pest. *Virochola isocrates* is the name of this insect. A spray with calcium arsenate controls the insect.

Posterior spiracles The larva of housefly has these in a pair and are D-shaped and these lie in the dorsal side of the 12th trunk segment.

Potato tuber moth Same as for vegetables.

Predaceous insects These are beneficial because they devour large number of other insects that are injurious. Many insects are scavenger type and vast quantities of dead animals and vegetable materials are eaten by them; thus preventing decay and unpleasant smells. Flies are especially very effective since they lay enormous number of eggs and the larvae, which hatch from them are extremely veracious. The maggots of certain blow-flies are used in curing of wounds. When placed in wounds, they feed on dead tissues and excreta, a substance known as allantion which stimulates healing.

Prementum The middle segment of mentum is smaller and it followed by prementum. The prementum at its basal angle bears a palpiger, which carries a three jointed labial palp. The pipartite inner portion is made up of an inner glossa and outer para-glossa.

Preoral cavity In insects, both stomedeum and proctodeum have a cuticular lining, continuous with the exoskeleton. The mouth parts enclose between them an indeterminate cavity known as preoral cavity. It is bounded anteriorly by the upper lip or labrum, posteriorly by the labium and on the sides by the two mandibles, and two maxillae.

Pretarsus *See* legs of cockroach.

Prispiracular glands Anoily secretion from these glands in a culex larva, is believed to be helpful in the suspension of the larva to the surface film.

Proboscis See mouth parts of the mosquito (insects). In the make mosquitoes, the palps are as along as the proboscis, but in the female the palps are much shorter. The proboscis is best developed in the female and consists of six lancer-like processes, laying very close together in a grooves sheath. A forward extension of the head.

***Procontarinia matteiana* Kieff and others** *See* Leaf gall makers of Mango.

***Prodenia litura* F** *See* Tobacco caterpillar of Tomato. *See* Leaf eaters of Colocasia.

Prolegs *See* legs of butterflies and legs of cockroach.

Propolis The material used by the builder bees for repairs, making block intruders of various kinds and sizes, of the comb which is different from wax and is supposed to be made by resin and other plant secretions.

Prothorax *See* thorax.

Protura In this order of insecta, small insects which lack antennae and compound eyes, have eleven segments of abdomen, occur in decaying matter in damp places, *e.g.,* Acerantulus.

Proventriculus This is situated at the posterior end of the insect crop and is pear shaped.

***Psara bipunctalis* F** *See* Leaf caterpillar of Brinjal.

***Pseudaulacaspis pentagona* Targ** *See* scale of Peach.

Pseudococcus bromeliae B *See* Mealy bug of Pineapple.

***Pseudococcus citri* R** *See* mealy bug of Pineapple.

***Pseudococcus citri* R** *See* mealy bug of Cocao.

***Pseudococcus virgatus* C** *See* Mealy bug of Custard Apple.

***Pseudococcus virgatus* F** *See* Mealy bug of Tomato.

***Pseudococcus lilacinus* Fab** *See* Mealy bug of Sappota.

Pseudopenis *See* titilator.

Psocoptera This order includes small insects with extremely long antennae and biting mouth parts. Males may not have wings but when present they have very much reduced venation, *e.g.*, book lice.

Psychodidae Sand flies. *See* Animal Science.

Psylla A sucking insect of Citrus.

Pterygota It is subclass of Insecta, usually have wings, the abdomen bears no appendages for locomotion, malpighian tubules are present, mouthparts hang free of head and metamorphosis is complete or incomplete.

Ptilium A bladder like structure which is situated on the front part of the imago of housefly. This is a flexible sac which is filled with blood. By forcing more flood into it, considerable force can be exerted. The fly uses this mechanism to knock out the anterior end of the pupal case very much in the manner of alid.

Ptilinial suture In housefly, once the fly is free from the imaginal case, the ptilium can be withdrawn into head in a contracted condition but a W-shaped mark, the ptilinial suture, persists on the head. Thus from legless, headless, and ligh-shunning maggots, living in the worst of filth, active sun-living aerial insects are formed. Also *See* metamorphosis of housefly.

Pulex ititans The human fleas are small wingless, laterally compressed insects which are parasitic only in their adult stage when they live in the feather or hair and such blood of birds and mammals. Their larvae are worm-like and live in the dust and dirt which collects in cracks, under carpets and in folds of upholstery.

Pulse beetles B. *Alnocallosus* Pic.

Pulse beetles *Bruchus analis* Fabr. Chocolate coloured beetles whose adult and grubs damage various kinds of pulses; eggs are laid on grains in store or on green pods in fields; grubs bore into grains or pods and feed inside for two to three weeks; pupal stage lasts for four or five days in summer and longer in winter; beetle (1) breeds only in stores and beetle (2) only in fields, others do so in both.

Pulse beetles *Laria affinis* Froel, *L. pisorum* Linn, *Pachymerus chinensis* Linn, *B. phaseoli* Gyll and *P. quadrimaculatus* Fabr.

Pulvillae *See* lets of housefly.

***Pulvinaria psidii* Maskell** *See* Mealy scale of Guava.

Pumpkin caterpillar of cucurbits *Margarania indica* S. A bright green caterpillar that feeds on all cucurbits; pupates in cocoon in leaf fold; the moth has white wings with dark brown marginal patches. Destroy caterpillars by handpicking. Dust with 5 per cent BHC or 1 per cent lindane @ 15 to 20 lb. per acre.

Pupa The third stage in the life cycle of insects that have complete metamorphosis, which follows the larval stage. During the pupal stage the insect does not feed and is usually immobile and internally it undergoes complete reorganisation of its structure. At the end of the pupal stage, the insect undergoes its final moult and the imago emerges. The pupa is often formed in a cocoon or made inconspicuous in some other way, as in the chrysalis of butterflies and moths. Mosquito pupa fare fully mobile.

Pupa of housefly The full grown larva, after a total larval period of 6 to 8 days, travel to a less moist and dark situation where it rests for a short time and then

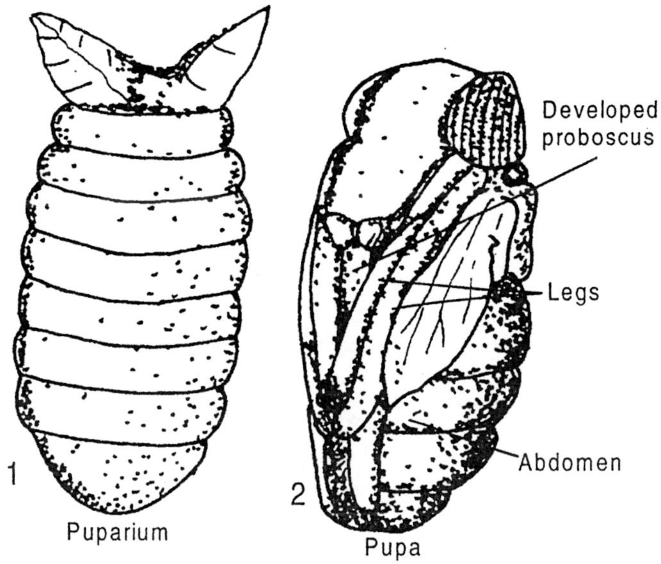

1: Empty puparium; 2: Pupa removed from puparium

changes to pupa. To begin with contraction of body takes place and the anterior segments are withdrawn. Consequently it becomes cylindrical and its anterior and postenionds become evenly rounded.

Puparium In the housefly, the last larval skin, instead of being shed, is retains and hardens to form a waterproof and air-tight ovoid pupal case of puparium, which merely serves as an outer proper protective covering for the larvae itself.

Pupation The larva of butterflies always pupates in some dry place. It hangs vertically with its head uppermost. Its body shortens and swells in the thoracic region, splitting its last larval skin or cuticle down the back after two or three days. By the rhythmic contraction of its body, it pushes off its old cuticle. On the last, segment is formed a group of hooks, the cremaster, and when the cuticle is shed the hooks anchor into a pad of silk that the caterpillar has previously spun. When this activity completed, the larva undergoes a metamorphosis or change into a resting pupa or chrysalis with a tough cuticle in which the larva passes through a series of changes.

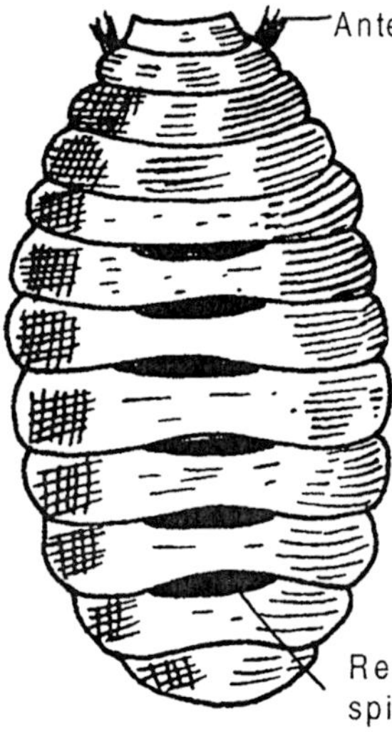

Pupa of Housefly

Pupal case of housefly In the pupa, the colour changes from creamy yellow to rich dark brown. The last larval skin, instead of being shed, is retained and hardens to form a water-proof and airtight ovoid pupal case.

Puparium of housefly Same as pupal case. It early serves as an outer proper protective covering for the larva itself. The pupa measures 6 mm in length and has no mouth and anus. The two anterior spiracles are situated at the anterior end while the posterior spiracles form button-shaped promonences at the evenly round posterior extremity.

Pupal stage of silk moth The silk thread is woven around the body of the caterpillar to form a covering called a cocoon, in which lies the animal in a dormant stage. This is pupal stage. *See* metamorphosis of silk moth.

Pyralis farinalis **Linn** *See* Meal snout moth.

Pyrilla perpusilla **Wlk** *See* Leaf hopper of Sugarcane.

Q

Quadraspidiotus perniciosus **Comstock** *See* San Jose Scale of Apple.

Quassia chips A decoction made from chips of quassia, woon has also been used and about as effective as nicotine sulphate; but expensive and more troublesome, it is not much used.

Queen-cells *See* honeybee.

R

Rasping sucking mouthparts found in honeybees.

Rats Once they gain entrance into stores and godowns, rate live and continually breed in them; they may also frequent the place from outside where they may have their burrows or hiding places; they spoil a great deal more or stored produced than they eat. These are the common post of farm crops and also of fruit trees by eating the roots. They may be poisoned by strychnine or some other rodenticide. Rats should be prevented from entering godowns at all costs. Anti-rat campaigns should be periodically organised and rats exterminated by (*i*) trapping them in rat traps, (*ii*) fumigating their burrows with calcium cyanide, (*iii*) baiting them with poison baits containing barium carbonate or zinc phosphide or strychnine hydrochloride. Poison baiting is not recommended inside the godown. But this operation may be carried out over as big an area around stores and godowns as possible wherever necessary, only warfrain baits may be used inside the godowns.

Receptor organs in insects Insects have sense of smell and taste, and are sensitive to light, sound, touch and changes of temperature. The receptor are isolated or collective modifications of epidermal cells to form sensillae, which are found situated in the various parts of the body.

Rectal papillae In the alimentary canal of cockroach, after ileum, there is a long and coiled colon which terminates in a broad rectum; the inner wall of the rectum has six longitudinal ridges projecting into the lumen. These are called rectal papillae which are said to be of help in the absorption of surplus of water and also perhaps in the distension of the rectum.

Rectum *See* rectal papillae.

Red ant of mango (*Mangifera indica* L.) *Oecophylla amaragdian* Fabr. Ants build leaf nests on the trees. Burn and destroy the nests or spray 0.25 per cent BHC on the nests.

Red borer of coffee (*Coffea arabica* L.) *Zeuzera coffeae* Niet. The caterpillars bore into the stem.

Red cotton bug *Desdercus koenicii*–damage many crops especially cotton and wheat.

Red gram plume moth of gram (*Cicer arietinum* L.) *Exelastis atomosa* W. A slender moth, with wings arranged in plumes; lays minute solitary eggs; caterpillar is greenish brown and bores into pods and feeds on seeds. Same as for gram caterpillar.

Red hairy caterpillar of sannhemp (*Crotalaria juncea* L.) *Amsacta moorei* Bulter. A caterpillar with long red hairs that feed on leaves; also cuts the seedlings at growing points; pupates in the soil; the moth is white with black dots on wing, and lays eggs on leaves or on soil. Trap the moth with light; plough deep after harvest. Dust with 10 per cent BHC @ 25 to 60 lb. per acre according to the stage of the caterpillar. Spray with 0.5 per cent BHC or 0.03 per cent endrin or parathion @ 40 to 50 gallons per acre.

Red mite A poultry disease. *See* Animal Science.

Red palm weevil of coconut (*Cocos nucifera* L.) *Rhynchophorus ferrugineus* Fb. A reddish brown weevil that lays eggs on the stem in a cavity scooped out by it; the grub feeds on soft tissues of the stem and kills the tree out-right. Close the holes with sand mixed with BHC or chlordane 5 per cent dust to prevent entry of weevils in the stem for egg-laying; inject 1 per cent Pyrenone (1 part pyrethrin with 10 parts piperonly butoxide) in the holes.

Red pumpkin beetles of Cucurbits *Aulacophora foveicollis* Fb. (A intermedia Jac). A red beetle that does extensive damage to all cucurbits; its grub bores into roots and fruits that lie on ground; pupates in the soil. Collect and burn all creepers, grasses, that provide shelter to beetles; plough deep after harvest to destroy grubs; in case of severe infestation, spray with 0.1 per cent lindane @ 40 to 60 gallons per acre, dust with 3 to 5 per cent DDT or BHC @ 2.5 per cent or 1 per cent lindane, @ 15–20 lb. per acre.

Red scale *See* scale insects.

Red spiders of tea (*Camellia sinensis* Kuntze) *Tetranychus bioculatus wod* Mason, (*Oligonychus coffeae* Niet). Nymphs and adults feed on the plant sap. Spray with lime sulphur wash (1 : 2 : 16) or 10–24 per cent aramite @ 60–80 gallons per acre.

Reproductive system in insects These sexes are separate *i.e.* the male and female germ cells mature in separate individuals. The reproductive system of both male and female can be divided into internal and external genetalia.

Repugnatorial glands Also called stink glands. The hinder segments of the cockroach are hidden so that the eighth and ninth terga of the female and male both, are largely hidden by the seventh tergum. This is a pair of repugnatorial or stink glands in the membrane between the fifth and sixth terga. Their secretion is believed to be deterrent to most enemies. In cockroach, there is a pair of these glands in the membrane between the fifth and sixth terga. Their secretion is said to be deterrent to most enemies.

Respiratory siphon The eighth segment of abdomen of Culex mosquito carries a long conical respiratory siphon which the larva must thrust into the air at intervals to breathe. *See* abdomen of culex.

Respiratory system in insects In the terrestrial insects like cockroach, there is a well developed system of air tubes or tracheae by which air is brought in direct relationship with the organs and tissues. The tracheae receive air through ten pairs of spiracles or stigmata, which are situated along the two sides of the body of the insect–two pairs in the thorax, one between

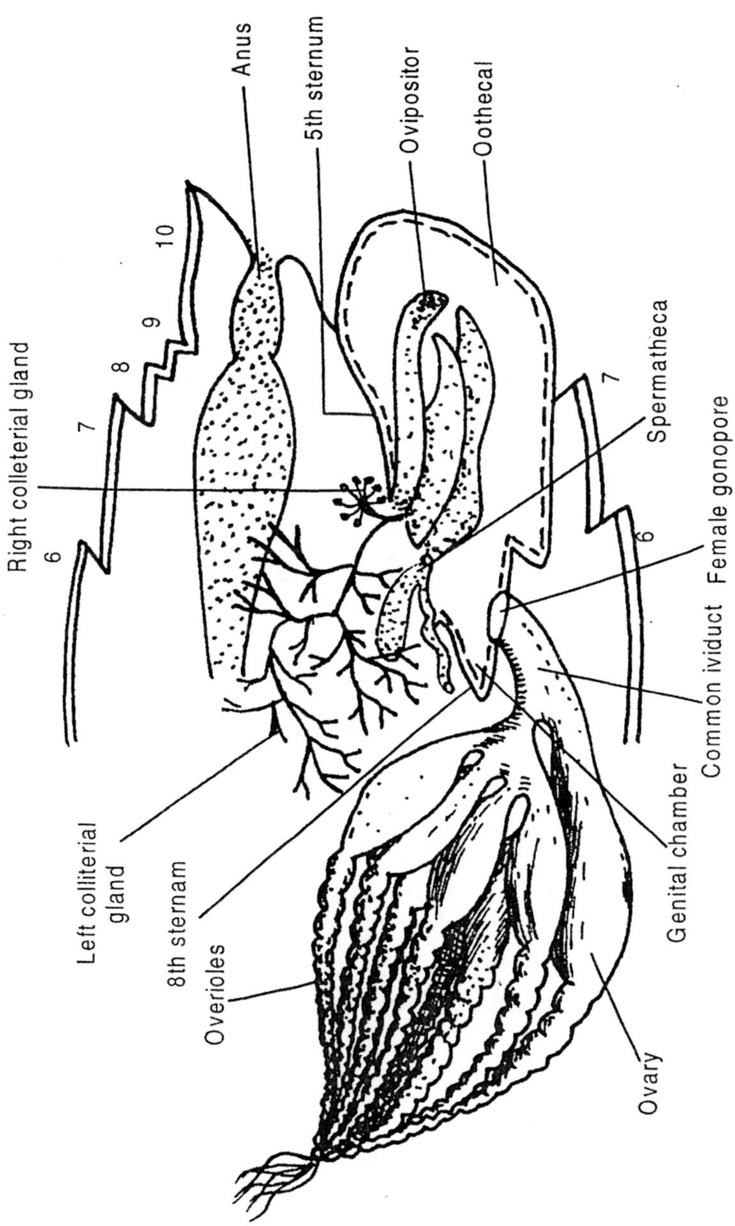

Reproductive organs of female Periplanata

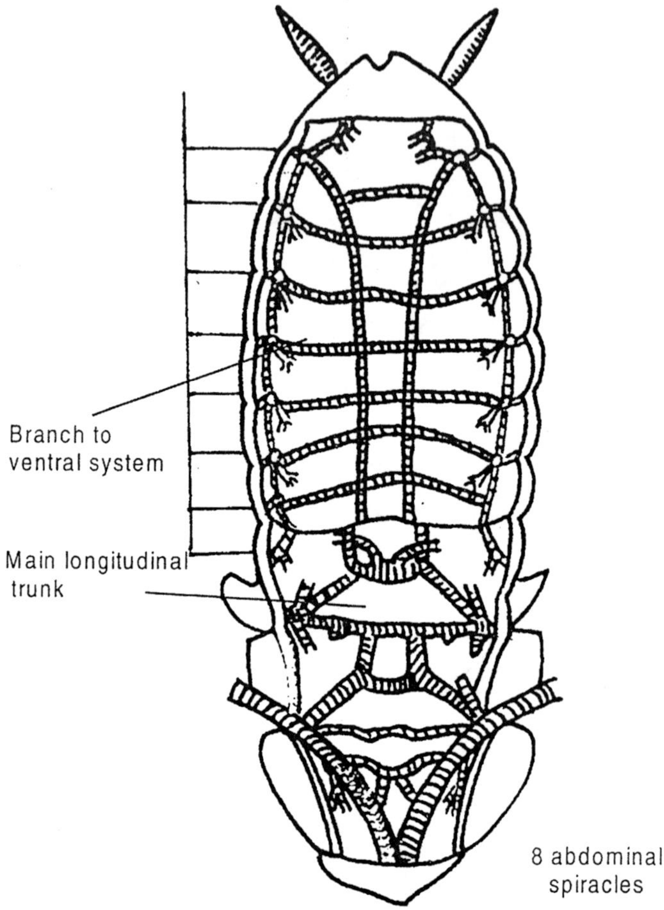

Arrangement of the main tracheae in the body of cockroach

prothorax and mesothorax and one pair between mesothorax and metathorax and eighth in the abdomen on each side of the first eight segments. It is by branched cuticle-lined tracheae that carry oxygen directly to the tissues from the paired apertures-spiracles or stigmata, which lie on the sides of thorax and abdomen.

Respiratory trumpets in a culex pupa The pupa is just as well provided with air tubes or tracheae as the larva but there are no respiratory siphon on the abdomen. Instead of this, there are two little respiratory trumpets which are situated on the back of the cephalo-thorax. They perform the same function as the siphon of the larva.

Resurgitation *See* mouthparts of honeybee.

Retinal cells *See* vitrelle.

Retinulae Same as retinal cells.

Rhabdom *See* arhabdomere.

Rhabdomere In insect eye, beneath the vitrellae is another group of seven cells, the retinal cells or retinulae, the inner borders of which become refractive each forming a rhabdomere. The group of rhabdomere forms the inner refractive body–the rhabdom.

Rhinoceros beetle of coconut (*Cocos nucifera* L.) *Oryctes rhinocerus* L. A stout black beetle with a pointed horn that borers into unopened fronds going right up to the growing point; the beetle lays eggs in manure heaps and decaying vegetable matter; the grub pupates in the soil, sometimes very deep. Hook out beetles from affected crowns; spray 0.01 per cent BHC on breeding pits; fill the leaf axils with 5 per cent BHC or chlordane dust mixed with equal quantity of sand or saw dust.

***Rhipiphorothrips cruentatus* M** *See* Thrips of Grape-vine.

***Rhizopertha dominica* Fabr** *See* Lesser grain borer.

Rhopalosiphum (aphis) maidis **(Fitch)** *See* Aphid of Maize.

Rhynchocoris humeralis **Thunberg** *See* Citrus green bug of Citrus Crops.

Rhynchophorus ferrugineus **Fb** *See* Red palm weevil of Coconut.

Rice grasshopper of rice (*Oryza sativa* L.) *Hieroglyphus banian* F. Nymphs as well as adults devour leaves and tender grains; usually active from September to December. Plough deep after harvest; dust with 5 to 10 per cent BHC at 20 lb. per acre. Spray with 0.02 per cent aldrin at 60–80 gallons per acre.

Rice hispa of rice (*Oryza sativa* L.) *Hispa armigera* Ol. Small blue-black covered with spines; grubs make long winding tunnels in leaves; pest more harmful to plants in seed-bed. Clean cultivation; dust with 5 per cent BHC at 15 lb. per acre. Spray with 0.25 per cent DDT @ 60 to 80 gallons per acre.

Rice moth, *Corcyra cephalonica* Stainton A dark grey moth that lays eggs anywhere in stores and godowns; creamy white caterpillars hatch out in about five days and feed on broken grains and subsequently bore holes in with bits of grains, grass, etc., in which they live; caterpillar stage lasts for 23 to 35 days and pupation about ten days; also attacks oil-seeds, dry, fruits, etc. Grain for storage should not contain more that 10 per cent moisture; store only one kind of grain in a godown; bags should not touch walls; ceilings or floors, and should be properly stacked, preferably on dunnage, by keeping suitable alleyways between stacks. Screen the grain before filling in bags. Gunny bags should be clean and free of insects; they may be fumigated if necessary before use. Dunnage material should also be free of insect and other pests, and should not have cracks or crevices in them. Vehicles used for transporting grain should be thoroughly cleaned and washed with water containing 4 per cent BHC regularly.

Rice weevil *Sitophilus oryzae* Linn. A small, reddish brown or black weevil with snout; adults and grubs feed on all cereals and millets; eggs are laid in excavations in the grain and the grub bores into the kernel where it feeds for three to four weeks, four to five generations in a year. The store or godown should be well ventilated and proof against dampness, insects and rats; walls ceiling and floor should be plain and smooth, without cracks or crevices; floor and walls up to six feet height should be painted.

Rickettsiae Obligate intracellular parasitic bacteria found in certain arthropods. They are not pathogenic to the arthropod, but if an infected animal bites a mammal, severe infection can result. Diseases caused by rickettsiae include typhus, psittacosis, and Rocky mountain spotted fever.

Rodenticide Chemicals used to kill rats pests *e.g.* zinc phosphide.

Rostrum Same as basi proboscis in mouthparts of houseflies.

Root borer of apple (*Pyrus malus* L.) *Dorysthemes hugelii* Redt and *Lophosternus bugalii.* A grub that feeds exclusively on thick root; the beetle is shiny black and lays oval eggs in the soil; sandy soil favours the growth of grubs. Remove and destroy grubs from roots of affected trees, avoid dry sandy soils for apple plantation and apply well rotten manure to the plants.

Root borter of sugarcane (*Saccharum officinarum* L.) *Emmalocera depressella* Swinh. A pale with caterpillar that damages the cane in early stages before nodes are formed; central shoot dries up; the moth lays eggs singly on the underside of leaves. Stop rationing; dig out and destroy stubbles after harvest.

Rural development Agro-processing and marketing has to develop a major activity which would provide linkages with industry and trade to enhance opportunities for rural employment.

S

Saccharicoccus sacchari **(Ckll.)** *See* Mealy bug of Sugarcane.

Safflower aphid of safflower (*Carthamus linctorium* L.) *Macrosiphum sp.* A black aphid that infests twigs and leaves. Same as for groundnut aphid.

Safflower caterpillar of safflower (*Carthamus linctorium* L.) *Perigoea copensis* G. A green caterpillar with purple markings that feeds on leaves. Remove larvae by handpicking. Dust with 5 per cent BHC @ 15 to 20 lb. per acre.

Safflower fly of safflower (*Carthamus linctorium* L.) is attacked by *Acanthiophilus helianthi* Rossi. A light green fly whose maggots damage buds and flowers. Trap flies in essential oil attractants.

Saissetia hemisphaerica **T** *See* Scale insects of Coffee.

Saliva It is the salivary secretion and contains amylase, invertase, maltase and lactase; which helps in digesting the food of organisms. With the help of this starch, sugars, maltose and lactose are digested to some extent.

Salivary glands of cockroach Associated with the alimentary canal is found a pair of salivary glands. Each gland consists of a bipartite glandular portion and a sac-like reservoir or receptacle. The two ducts from the glandular portions of each side unite to form a glandular duct. Salivary glands are situated in the thorax.

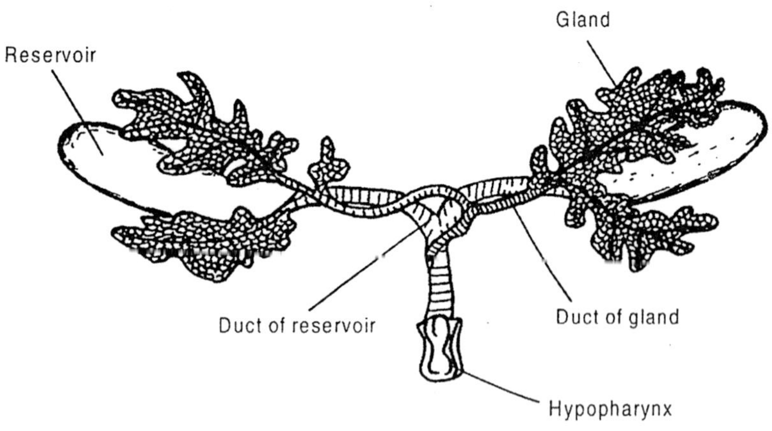

Salivary Glands of Cockroach

San jose scale *Aspiodiotus (Quadraspdiotus* or *Aonidiella) emiociosus.* Introduced in Kashmir in about 1910 where it threatened to wipe out the fruit industry. Now also found in Punjab, the NWF provinces and Uttar Pradesh. Spraying with diesel oil, fish oil and rosin kills practically all scale insects.

San jose scale of apple (*Pyrus malus* L.) *Quadraspidiotus pemiciosus* Comstock. A sucking insect on twigs, branches and fruits; attacks a large number of other plants. Spray dormant trees in winter with 4 to 6 per cent diesel oil emulsion @ 1–2 gallons per acre.

San jose scale of plum (*Prunus domestica* L.) Same as for apple.

San jose scale of pear (*Pyrus communis* L.) Same as for apple.

Saw-fly of mustard *Athalia proxima.* A serious insect pest of mustard crop.

Saw like ends *See* mouthparts of mosquito.

Saw toothed grain beetle *Oryzaephilus surinamensis* Linn. A slender brown beetle which does not fly though it has wings; both beetle and grub feed on flour and even grains with high moisture content, dry fruits, flours, meals, etc potatoes after harvest should not be kept exposed but kept covered suitably till carried to the godown. The tubers should be placed on and covered with thin layers of clean dry sand dusted with 5 per cent DDT. Inspect the tubers frequently and remove and burn all diseased and infested potatoes. Infested tubers may be fumigated with petrol or methyl bromide.

Scale insects of orange *Sphceloma tawcetti* is the causal organism. Some time mistaken for the canker, because of its similar appearances. Lesions are found on leaves and affected parts. To reduce the attack of this insect, plants may be covered with copper fungicide sprays. The red scale is a serious pest of Citrus. It is *Hoinidiella auranti*. The yellow scale is A. orientalis is another pest of Citrus.

Scale insects of banana Three types of scale insects *Aonidella aurantii, A. knemantalis* and *Aspidiotus destructer* are the main scale insects of a banana crop.

Scale insects of coffee (*Coffea arabica* L.) *Lecanium viride* G. Red and green scale insects found on tender shoots and leaves. Prune infested shoots; spray with fish oil rosin or 0.02 per cent parathion or 0–1 per cent HETP @ 2–3 gallons per tree.

Scale insects of coffee (*Coffea arabica* L.) *Saissetia hemisphaerica* T.

Scale insects of guava *Pulvinaria psidii.*

Scale insects of litchi (*Litchi senensis*) *Parlatoria pseudopyrii.*

Scaly leg of poultry A spider caused disease of poultry. *See* Animal Science.

Scape The broad based basal segment of an insect antennae.

***Sceledonta strigicollis* Mots** *See* Flea beetle of Grape-vine.

***Schizaphis (Taxoptera) graminum* (R.)** *See* Wheat aphid of Wheat.

Schoenobius incertellas *See* stem borer of paddy.

***Scirpophaga nivella* Fabr** *See* Top shoot borer of Sugarcane.

***Scirpophaga nivella* Fabr** *See* Chilli thrips of Chilli.

Screw-worm True screw-worms seldom get through the unbroken skit but will penetrate moist pockets like the prepuce of a gelding or the dimple in front of a cow's udder. They are not found in cold blooded animals, such as turtles, snakes and lizards (reptiles). Manmade wounds resulting from branding, castrating, dehorning and docking afford a breeding ground for this parasite. The screw-worm fly is bluish green in colour with three dark stripes on the back and reddish or orange colour below the eyes. When these worms have reached their full growth, they assume a pinkish colour, leave the wound, and drop to the ground where they dig beneath the surface of the soil and undergo a transformation to the hard-skinned dark-brown motionless pupa. During this stage, the maggot changes to adult fly. Any control measure should be taken during winter season when the fly is slow in breeding and other activities. Treat the wounds with a dust 5 per cent D Caumaphos on wounds and surrounding area thoroughly. Repeat as necessary.

Second instar of housefly After a day or two of active feeding, the first instar larva sheds its skin and becomes the second instar. It is distinguished from the first by the possession of a pair of fan-shaped anterior spiracles.

Semilooper These two species *viz. Plucia orichalcea* and *P. nigoiligna* damage the wheat crop. Control measures are the same as for army worms.

Sense of audition Auditory sensillae are believed to be present on the anal cerci. Scattered sensillae in the palps, legs and anal cerci probably help to detect mechanical strains in the chitin of the exoskeleton.

Sense of smell in insects This is well developed. Organ smells are present on the antennae and also on the palps.

Sense of taste Organs for this are present on the tips of the maxillary and labial palps.

Sensilla It consists of receptor or sensory cells often in relation with the cutinlar development such as bristle or hair. *See* receptor organs of insects.

Sensory system The central nervous system of cockroach is of usual annulate type, resembles in its main features that of cockroach. It consists of a supraoesophageal commisures or cerebral ganglia, subeosophageal commisures, in the head and a double ganglionated ventral nerve cord in the thorax and abdomen.

Sericulture The rearing of silkworms to produce raw silk. This is a labour intensive form of farming using cheap skilled workers. Fresh mulberry leaves are fed to the silkworms and the rhread unravelled from the cocoons on small spinning machines. The main silk-producing countires are China, Korea, Japan and Italy. Also *See* Crop Science.

Sesame bud fly *Desineur asasanis* Full fed larvae cut a hole in the bud, drops out, pupates in the soil 5-10 cms deep, usually pupation occurs in a silken cocoon.

Sesame gall fly *Asphondelia sesami* Maggots of this fly are parasites by *Euroritomid dentipictus.*

Sesame leaf hopper *Orosius altionetus.*

Sesame wheat fly *Burmicia takaci.*

***Sesamia inferens* Walk** *See* borer of Wheat and Pink borer of Maize.

Sesbamia leaf spot caused by *Cercospora sesbanis.*

Sexes in Arthropods Sexes are usually separate in Arthropods and the male and female are easily distinguishable.

Sharp claws *See* legs of cockroach.

Shoot borer of Jack fruit (*Artocarpus integrifolius* Auth.) *Margaronia caesalis* W. A reddish brown caterpillar, turning green when old, that bores into shoots, buds and fruits. Destroy attacked shoots, buds or fruits fumigate the borer holes with petrol of ED/CT mixture.

Shoot borer of apple (*Pyrus malus* L.) *Alcides mali* Marshal. A weevil that lays eggs in tender shoots; the grub bores into the shoot causing a gall-like swelling and makes a tunnel down the stem. Remove and destroy swollen twigs; collect and destroy weevils.

Shoot flies of maize *Allerigona soccata, A. nagavi* and *A. orientalis.* To control, apply at the time of sowing in furrows phorate 10G or Chloroferon 3G @ 1 gm/m row or controlled by release of laboratory prepared 10G (Polyethelene glycol–PEC 6000) at the rate of 0.5g/m row or furiothiocarb 10G (Promet 10G) or quinolphus 1.5 Q @ 2.5 kg/ha. Or deed treatment by carbosultans 25 Std. @ 40 g/kg seed.

Shoot fly *Alterigona haqvii.* This adult is like a housefly. This is said to be parasitised by some species of *Tetrasticus* and *Neochygotharis* among *T. nayanitanus* is important.

Shoot hole borers of coffee (*Coffea arabica* L.) *Xyleborus morstatti.* Same as for shot hole borers of tea.

Shoot hole borers of coffee (*Coffea arabica* L.) X. *compactus E.*

Shoot webber *Orthaga exvinacea.* The caterpillars cause considerable damage to the leaves of the mango trees.

They can be controlled by hand picking the infested clusters or by spraying with calcium arsenate.

Silk glands By the time the work has reached maturity, the silk worm is about 8 cms long. Running along it the sides of its body, just inside, are two bags of sacs. These are silk glands. As the larva attains it matures size *i.e.* before the fifth moult, these silk glands become filled with a sticky substance which hardens to solid when exposed to air. *See* metamorphosis of silk moth.

Silk moth It is about 25 mm long and its body is divided into three portion namely the head, thorax and

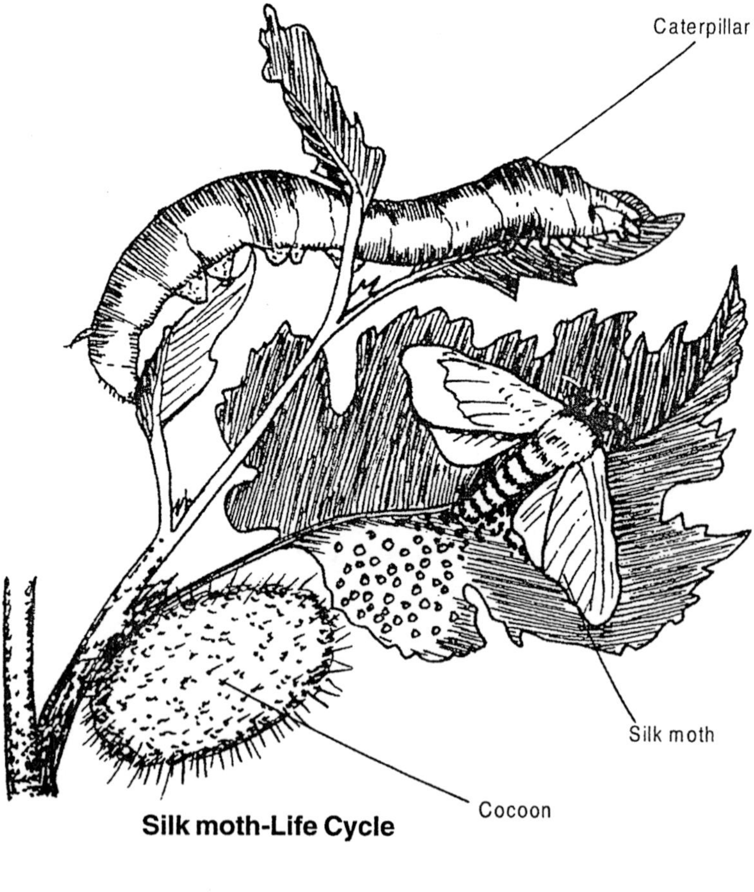

Silk moth-Life Cycle

abdomen. They bear eyes, a pair of antennae and a mouth. Thorax bears three pairs of legs and two pairs of wings. Wings are about 25 mm long and are cream coloured, lined with stripes of light brown colour.

Silver fish (*Lepisma saccharina* Lin) An glistening grey wingless insect with its body tapering towards both ends, two long feelers on the head, and three long process at the opposite ends; infests dark, moist, warm places and feeds on flours, meals and other food products.

Simufiicae The buffalo gnats. *See* Animal Science.

Siphonaptera *See* Aphanipetera. It includes small wingless insects which are flattened on the sides (laterally), piercing and sucking mouth parts are present. Head small, no compound eyes, and legs adapted for leaping Ectoparasites of mammals, birds and others, *e.g.*, fleas.

Siphunculata It includes insects which are external parasites on mammals. With their piercing mouthparts, they suck blood. *e.g.*, sucking lice.

***Sitodrepa panicea* Linn** *See Sitodrepa panicea* Linn.

***Stiophilus oryzae* Linn** *See* Rice weevil.

Snake gourd semilooper of cucurbits *Plusia peponis (F.).* A whitish green semi looper that cuts the leaf from the edge and makes a nest with the cut portions on the leaf; pupates inside the nest. Same as for pumpkin caterpillar.

Solitary locust *See* phases of locusts.

Somites Same as segments of cockroach. In the embryo, there are two somites but in the adult these are ten. In housefly in male, there are eight segments but in female there are nine. In mosquito abdomen, there are nine segments. Each somite has a dorsal tergum.

Spermetheca In cockroach, is made of two slender, spirally coiled tubules of unequal size united to form a short stalk which opens into the roof of the vagina.

Spermetophore In cockroach, before copulation the sperms, which are present in the seminal vesicle, stick together to form a small capsule called spermatophore.

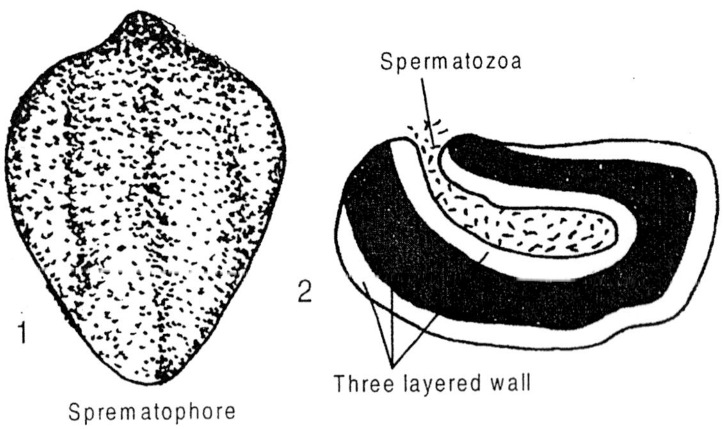

1: Spermatophore; 2: The same in cross section

Sphenoptera gossypi **G** *See* Cotton stem borer of Cotton.

Sphenoptera perroteti **G** *See* Stem borer of Groundnut.

Sphingid moths These are the three species of small insects damaging grape vine *Theretra alecta, Caleris lineata* and *Herse conconvuli.*

Spiniferous pads On the ventral side of the housefly, at the anterior edge of each of the 6th to 12th trunk segments, these pads exist in a semilunar shape covered with fine recurved spines. These are locomotary in function; are used by the larva in cooperation with the mandibular sclerite in wriggling both back and forward.

Spinneret Before the fifth mould, the silk worm goes without food for a long time, and usually hides itself in a comer. Then the sticky fluid is forced out of the silk glands through a common pore called spinneret, which is situated on the under lip of the caterpillar. As the sticky material is forced out into the air, it forms a long solid thin strand of silk. One silk worm can eject about a thousand yards of silk thread in this manner.

A paired appendage on the abdomen of a spider for spinning silk, used for making its web or egg cocoons or for binding prey caught on the web. The silk is secreted as a liquid by silk glands but hardens as it passes out through the spinnerets.

Spiracles (1)–An opening leading to a trachea in an insect, isonod, centipede, millipede or arachnid. Spiracles occur in pairs on each side of the body. In adult insects, one pair occurs on each of the posterior, two thoracic segments and the anterior eight abdominal segments. The arrangement in larval and pupal stages of insects may vary. (2)–The opening from the gill chambers (antria) of the frog tadpole, on the left side only. (3)–The anterior gill cleft of cartilaginous fish, usually much reduced.

Spirochaetosis A tic disease of young fowl. *See* Animal Science.

Spondoptera mauritia **Boisd** *See* swaming caterpillar of paddy.

Sponging type of mouthparts *See* mouthparts of housefly.

Sponge-like organs *See* mouthparts of housefly.

Spoon-like bowl *See* mouthparts of honeybee.

Spotted fever tick Induces fever in animals and man.

Spotted Fever Tick (Dermanyssur)

Stalk borer of maize Damage caused by insect *Chilo partellus.* Can be controlled by foliar application of 0.05 Lindane 3.5 E.C. 10–15 days after germination.

Stem borer of mango *Bactocera rufomaculata.* It is a grub of this insect having long large horned beetle.

Stem borer of apple (*Pyrus malus* L.) *Zeuzera* sp. A caterpillar that bores into branches and stems. Remove and destroy twigs while pruning; fumigate with petrol or EDCT mixture in the case of big branches.

Stem borer of apple (*Pyrus malus* L.) *Linda nigroscutata* Fairm. A beetle whose grub bores into twigs in which it makes holes to thrust our 'pellets of frass'; the beetle lays eggs under the bark. Remove and destroy infested twigs.

Stem borer of banana (*Musa paradisiaca* L.) *Odoiporus longicollis* Ol. A weevil grub bores into stem and kills the plant; weevil lives for 1-2 years. Uproot and burn infested plants.

Stem borer of groundnut (*Arachis hypogaea* L.) *Sphenoptera perroteti* G. A shining dark colour beetle whose larva burrows into stem and pupates in it. Remove and destroy infested plants.

Stem borer of paddy (*Oryza sativa*) Pest active between March and November; caterpillars bore into stem; central shoot withers and produces dead heart; plough up and destroy the stubbles after harvest. Collect and destroy egg clusters in nursery plants. Dip seedlings in 0.1 per cent DDT suspension before transplanting. Spray the young and the mature crop with 0.025 per cent parathion or 0.08 per cent endrin in the field @ 60–80 gallons/acre. The borer is *Schoenobius incertellas* Wlk.

Stem borer of sugarcane (*Saccharum officinarum* L.) Gurdaspur or Dehradun borer, *bissetia steniellus* Hamps. Plough up and destroy the stubbles; remove infested tops once every week for three months from the time when infestation is first noticed.

Stem borer of sugarcane (*Saccharum officinarum* L.) *Chilotraea influscatellus* Sn. C. *nuricilia* Dudg. A caterpillar of dirty cream colour that bores in the stem and feeds inside; the month is greyish brown and lays eggs in clusters on underside of leaves; central shoot becomes white and dries up; and the cane has a large number of holes. Plant sets free from borer; stop ratooning; give two light earthings during early stages of crops; use light traps; remove watershoots.

Stem borer of sugarcane (*Saccharum officinarum* L.) *Chilo tumidicostalis* Hamps. *Procerss indicus Kapur Raphimatopus ablutella Zell.* Spray with 0.1 per cent endrin or 0.32 per cent DDT @ 100–150 gallons per acre; mixing 10 per cent BHC dust in the soil @ 50 lb. per acre while plant sets, against early short borer.

Stem borer of Citrus *Iiderbela quadrinotata, Stromatum barbatum, Monghammus vereteegi* and *Chelidomum cinctum* are the various borers that attack citrus stems. Can be controlled by breaking away the affected dead twigs. Painting the galleries with aris green, thursting a wire into the hole or injecting a little petrol, or a mixture of two parts of chloroform to one part creosote controls the insect.

Stem borer of wheat (*Triticum sativum* Lam.) *Sesamia inferens* Wal. A pinkish brown caterpillar with cylindrical body; bores into stem and kills central shoots; the moth is straw coloured and lays eggs in clusters inside leaf sheath. Spray with 0.03 per cent endrin @ 60 to 80 gallons per acre. Dust with 5 per cent BHC or DDT @ 15 to 20 lb. per acre.

Stem borer beetle of soybean (*Glycine max* Merr.) A small pale brown longicorn beetle whose grub bores through the stem. Dust with 5 per cent BHC @ 1 lb. per acre.

Stem fly of cowpea (*Vigna sinensis* Savi) *Agromyza phaseoli* C. A small blue black fly that lays eggs on stem of young plants; its pale yellow maggot bores into stem. Same as for gram pod fly.

Sternal haemocoel *See* ventral diaphragm.

Sternum in cockroach The first sternum is rudimentary. In the male, nine sterna are visible externally but in the female cockroach only seven are visible.

Sthenias grisator fabr See Vine girdler beetle of Grape-vine.

Stigmata Same as spiracles.

Stink glands *See* refugnatorial glands.

Stipes *See* maxillae.

Stock fly Same as stable fly and horn fly of cattle.

Stomach poison Most stomach poisons are arsenic compounds. The most commonly used is lead arsenate, which is marketed as a greying white powder or paste. It is dangerous poison and should be handled with great care. These stomach poisons are used against insect pests which cause injury to the plants or trees by biting and chewing their various parts. Other stomach poisons are Calcium arsenate, Lead chromate, Sodium Fluosilicate, hellebore (plang), and the last one is also used as contact.

Stomodeal valve In the hinder more tapering part of the proventriculus behind the thick provenctricular plates, in cockroach, there is another circule of six lightly cutinized thin proventricular plates which are followed by six cushion like pads covered with cuticular hairs which are all directed backwards. Beyond these, cushion is the region of the stomodeal valve which projects into the cavity of the anterior of the mesenteron.

Stomodeum *See* alimentary canal of cockroach.

***Stomopteryx nberteria* M.** *See* Leaf miner of Groundnut and Soyabean.

Stridulation The production of sounds by some insects, usually males, by rubbing together parts of the body.

Grasshoppers stridulate by rubbing the hindlimbs against the forewing; crickets by rubbing forewings together. Stridulation plays an important part in attracting and stimulating a female during courtship.

***Somatium basrbatum* Fabr** *See* orange beetle borer of citrus crops.

***Strumeta ferrugineus* Fabr** *See* Fruit flies of Manog.

***Strumeta incisus* W** *See* Fruit flies of Mango.

***Strumeta zonatus* Saund** *See* Fruit flies of Mango.

Strumentum The basal segment of the maxillae of insect, which is very broad from side to side.

Stylophyga orientalis This species of cockroach is of Asiatic origin. In the female of this insect, the female bears vestigial wings.

Suboesophageal ganglia It gives several nerves innervate the mandibles, maxillae and labium. In the thorax and abdomen, the nerves are segmentally arranged except that those arise from the last abdominal segment.

Suboesophageal ganglia in cockroach These are a pair of large ganglia lying between the submentum and the oesophgus.

Sucking insects of brinjal (*Solanum melongena* L.) *Phenacoccus insolitus* G. A mealy bug that attacks isolated old plants. Spray with 0.1 per cent malathion.

Sucking insects of brinjal (*Solanum melongena* L.) *Eurentins echinus* D. A lacewing dark brown bug that feeds on leaves which turn yellow and dry up. Dust with 5 per cent BHC @ 15 to 20 lb. per acre. Spray with 0.2 per cent BHC or 0.004 per cent Pyrocolloid @ 60 to 80 gallons per acre.

Sucking mouthparts These are formed in butterflies and moths for sucking nectar. These are tubular mouthparts.

Sucking tube *See* mouthparts of honeybee.

Sugarcane mite of sugarcane (*Saccharum officinarum* L.) *Oligonychus indicus* Hirst. Nymphs and adults feed on the sap of the plants. Spray with 0.25 per cent aramite or 0.025 per cent parthion or lime sulphur wash (1 : 2 : 16) @ 100 to 150 gallons per acre.

Suprapharyngeal ganglia The group of rabdomeres forms the inner refractive body known as rhabdom. From the inner ends of the retinal cells pass back nerve fibres which are connected to this ganglia. Surrounding each ommatidium are iris and retinal pigment cells in which the pigment is not retractable and therefore each ommatid is completely isolated optically, from the neighbours.

Supra-oesophageal ganglia in cockroach These are a pair of large closely opposed ganglia in the head and lie supported by an internal framework of chitin called tentorium.

Surface grasshopper of wheat (*Triticum sativumlam*) *Chrotogonus* spp. *See* under rice grasshopper. Dust with 5 per cent BHC @ 15–20 lb. per acre.

Swarming of bees See Flight to death.

Swarming caterpillar of paddy (*Oryza sativa*) *Spodoptera mauritia* Boisd. Caterpillars appear in big swarms in seed-beds; cause heavy depredation when cold weather is suddenly followed by a spell of warmth. Flood paddy fields or surrounded then with trenches having steep sides; dust with 5 per cent BHC at 15 lb./acre.

Swarming Same as flight to death in bees.

Sweet potato weevil of sweet potato (*Ipomoea batalas* Lam.) *Cyclas formicarius* F. A red and blue ant-like weevil that makes holes in stems and tubers; its pale yellow grub which has no feet bores into vines. Use insect-free vines for planting; remove infested vines and tubers; plant deep-rooting varieties; dust seed with 5 per cent DDT at 1 lb. for 8 to 10 md. of seed before sowing.

Sweet potato sphinx (*Ipomoea tatalas* Lam.) *Herse convoluvli* L. A stout long horn caterpillar that completely defoliates the plant; pupates in an earthen and destroy caterpillars; plough up infested fields after harvest; in case of severe infestation, spray with 0.25 per cent DDT @ 60–80 gallons per acre. Collect and destroy egg masses and caterpillars.

***Sylepta derogata* F.** *See* Cotton leaf-roller of Cotton.

Symbiotic fungus in mosquito The irritation produced after the insertion of the sharp stylets during mosquito bite, the subsequent suction is due to the symbiotic fungus. This fungus produces lot of carbon-dioxide, is said to cause irritation of the muscles while the ferment is said to increase the blood pressure.

Symphyta It is one of the major divisions of hymenoptra, and includes the primitive members. None are social insects. The best known forms are the wood-wasps which have their ovipositors for wood-boring.

Systemic insecticide Such insecticides which on application in proper dozes get into the system or body of the plant and when insects attack these plants, they are killed immediately. Such insecticides should be used with care when treating vegetables and the like crops because the effects of the insecticides are for a lengthy period and treated vegetable crops should be used only when the bad effects have completely attained a non dangerous stage for human consumption and even for animals, such insecticides are Dimecron, endosulpan etc.

T

Tabanideae Horse fly. *See* Animal Science.

Tamarind beetle, ***Caryoborus gonogra*** **Fab.** A grey beetle whose grubs bore into preserved tamarind fruits.

Tantorium It is a chitinous frame work supporting in the head of cockroach, the spuraoesophageal gangalia, which are a pair of large closely opposed ganglia.

Tar *See* borer.

Tarsomeres The insect tarsus is made of five movable parts known as tarsomeres. The terminal tarsomere is known as pretarsus: It ends in two lateral sharp claws and a small median hairy porous pad, the arolium which is an adhesive organ. The lower sides of the tarsomeres also have small adhesive pads known as pkantulae. The arolium and the plantilae enable the insects to walk upon smooth and vertical surfaces and even on window-panes ceilings and other similar places.

Tea looper of tea (*Camellia sinensis* Kunte) *Buzura suppressaria* G. Caterpillars defoliate the plant. Spray with 0.16 per cent DDT @ 60 to 80 gallons per acre.

Tea mosquito of tea (*Camellia sinensis* Kuntze) *Helopeltis antonii* S.H. *babriculasa* Bar. H. Chlivara Waterh. A slender sucking bug, mixed red, black and white in colour, that feeds on leaves and ender shoot which often curl up, fade and die. Collect bugs by hand nets and kill by dropping in kerosenised water; spray with 0.1 per cent DDT @ 60–80 gallons per acre.

Tea red slugs of tea (*Camellia sinensis* Kuntze) *Heterusia cingala* Moorre, *H. Magnifica* Butl. Caterpillars defoliate the plant. Spray with 0.16 per cent DDT @ 60–80 gallons per acre.

Tegmina The forewings of insects are called wing-covers or tegmina.

Tengum The plate, stiffened by chitin, that forms the protective dorsal covering to each segment of the thorax and abdomen of an insect. *See* also pleuron, sternum.

***Tenebroides mauritanicus* Linn** *See* Yellow meal worm.

Tent caterpillar of apple (*Pyrus malus* L.) *Clisiocampa indica* Walk. A caterpillar that is gregarious in habit in early stages feeds on leaves at night and lies during day in small tent-like structures of web; moth lays eggs in rows around small twigs in May from which caterpillars hatch out next spring. Remove and destroy egg rings during pruning; spray web tents with 0.5 per cent DDT @ 1 to 2 gallons per tree.

Tentorium *See* Supra-oesophageal ganglia in cockrbach.

Tergosternal muscles The respiration is also helped by the ventilation movements produced by the contraction and expansion of the abdomen. During contraction (in insects), the terga and sterna approach each other due to the contraction of the tergosternal muscles, in the vertical plane, when these muscles relax, they separate.

Tergum *See* Somite.

Termites, *Odontotermes* spp. Also known as white ants, are really not ants at all and belong to entirely different order. They attack particularly structures and articles of wood.

Termites Termites, like ants, live in colonies, with a well-organised social system, consisting of workers, soldiers, drones and queen; they attack growing plants *e.g.* sugarcane, wheat, etc. as well as dead organic matter like wood, gunny bags etc.

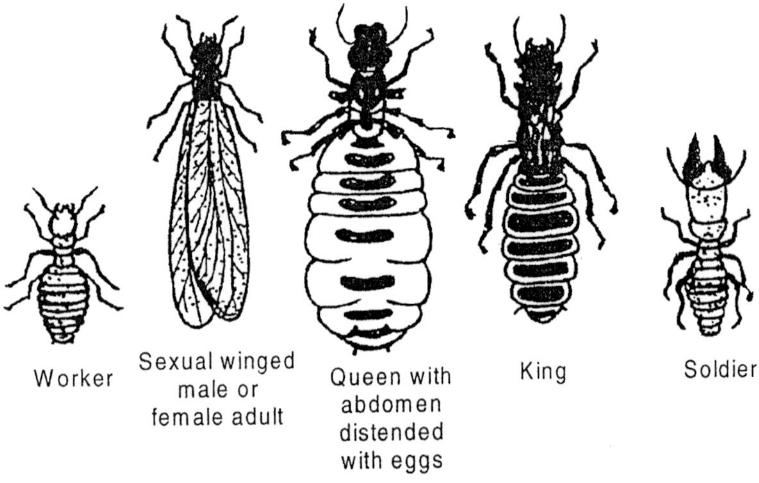

Caste of Termite

Termites of grapes *See* hymenoptera.

Termites of mango (*Mangifera indica* L.) *Neotermes* sp. *Odonototermes ogesi* Ram. Termites feed on the bark and underlying tissues. Mixing 5 per cent aldrin or chlordane dust in the soil around the tree or paint 5 per cent aldrin or chloridane emulsion round the stem upto a height of six feet from ground level.

Termites of sugarcane (*Saccharum officinarum* L.) *Odontotermes obesus* Rand O. *assumathi* Hlmgr. White ants that feed on seed or buds of planted setts. Adopts preventive measures; apply crude oil emulsion in irrigation water; repeat till plants are well established. Before planting dust furrows with 5 per cent BHC, aldrin or chlordane dust @ 30–40 lb. per acre.

Termites of sugarcane (*Saccharum officinarum* L.) *Erenotermes parapoxalis* Hlmgn.

Termites of sugarcane (*Saccharum officinarum* L.) *Trinevitermes heimi* Wasm and *Microtermes obesi* Hlmgr.

Termites of tea (*Camellia sinensis* Kuntze) and wheat *Microtermes* sp. *Microcrotermes* sp. Termites attack the dying and dead tissues of the tea plant. Mixing 5 per cent aldrin or dieldrin dust in the soil @ 20 lb. per acre.

Termites of wheat (*Triticum stivum* Lam.) *Microtermes obesi* (Anandi) H., *Odontotermes obesus.* Attacks young seedlings as well as grown up plants. Mix 5 per cent BHC, aldrin or chlordane dust in the soil @ 20–25 lb. per acre before sowing the crop.

Termite mound The nest, made from mud or plant debris, that houses a colony of termites (a tropical type of ant). Millions of individuals in a highly organised social system inhabit such mounds, which rise several metres from the ground surface; they appear to improve the structure of the soil but also cause some soil creep.

Terrain An area of land considered in terms of its physical characteristics, hence rough terrain or smooth terrain. Terrain analysis is based on the scientific classification of landscape units considering the characteristics of the soils, slope angles, geomorphic context, and processes of development.

***Tetranychus bioculatus* Wod Mason** *See* Red spiders of Tea.

Third instar of housefly In two days, the second instar larva moults and becomes the full grown third instar. It measures about 6 to 12 mm in length and due to the accumulation of considerable fat, it is creamy white in colour. The full grown larvae shun light and live just below the surface or rotting and fermenting organic debris, where most favourable conditions of temperature and moisture prevail.

Thorax of insects In insects, it is the portion to which the head is connected through a short narrow neck or carvix. Thorax is made up of three segments, prothorax, mesothorax and metathorax; each segment consists of a dorsal plate, the sternum joined to one another at the sides by a membrane, the pleurone. The tergum of the prothorax is so enlarged that it not only hides the neck completely but also overtaps part of the head and mesothorax. Each sternum bears a pair of legs. The mesothorax and metathorax bear each a pair of wings jointed to the anterior corners of the notum.

Thorax of mosquito The thorax is composed of three segments but it is impossible to see then all for one is very much larger than the other two. Each of them bears, however, a pair of jointed legs and high up on the sides of the middle segment of mesothorax are attached single pair of wings. Mosquito has only two wings and the second pair is represented by two club-shaped processes or halteres, situated on each side of metathorax.

Thrips Three species of thrips *Anaphothrips flavicincitus janglankabenori Haplothrips* spp. and Thrips damage many crops.

Thrips of grapes *Rhipiphorothrips cruentatus*, lacerate the leaves and suck the juice. Spray of dust Nicotine insecticide.

Thrips of grape-vine (*Vitis vinifera* L.) *Rhipiphorothrips cruentatus* M. They infect leaves and such sap. Spray with 0.05 per cent nicotine sulphate lindane or 0.16 per cent DDT @ 150–200 gallons per acre.

Thysanoptera It includes small insects with piercing mouth parts and short antennae. Both pairs of wings are similar with reduced venation, *e.g.*, thrips.

Thysanura In this order such insects are included which have chewing mouthparts, abdomen consisting of 11 segments, some of which carry appendages, *e.g.*, bristle tails and silver fish.

Ticks and mites Though these belong to class arachnida but due to their body structure, they are at times mistaken as insects. In these creatures, the body is rounded with no demarcations between the prosoma and opisthosoma. A number of species of ticks and mites infest our domestic animals. They are small in microscopic in size. The itch mite *Sarcoptes scabei* burrows in and irrates the human skin. The scabmite produces serious damages to the sheep and the chicken-mite causes unthriftiness in fowls. Ticks infest both domestic animals and man. Wherever they are present in large numbers, they make the host anaemic and may even cause its death. Also *See* Animal Science.

Tinea pellionella It is the case bearing clothes moth, is brown with three dark spots on each forewing.

Titilator The left phallomere is a complicated structure and has several parts arising from the common base on the extréme left, it has a long slender arm ending in a curved hoom known as titilator. Next to it lies the psuedopenis. Next to it lies the hook shaped asperate lobe.

Tobacco caterpillar of tomato (*Lycopersicon esculentum* Mill) *Prodenia litura* F. A stout dark brown caterpillar that damages leaves, shoots and fruits; the moth is dark with white markings on upper wings.

Tobacco borer beetle, *Lasioderma serricorne* Fabr. A dark brown beetle whose g rubs are very hairy and curled; both bore into cigarettes, cigars, dry tobacco and pulses; they also attack stored coriander, cumin, turmeric and ginger.

Top shoot borer of sugarcane (*Saccharum officinarum* L.) *Scirpophaga nivella* Fabr. A creamy yellow caterpillar that makes a tunnel in the leaf mid-rib and bores into the upper nodes; the white moth with silvery wings lay egg masses on the underside of leaves; central shoot dries up followed by production of side shoots. Collect egg masses frequently; remove and destroy top shoots infested with 1st and 2nd broods; use light traps.

Toxoptera *Citricidus* Fab. *See* citrus aphids of Citrus Crops.

Trachae *See* respiration in insects.

Tracheal capillaries in insects The spiracular aperture leads into a cavity, the tracheal chamber, from which the main tracheal trunks arise. From the two main trunks branches are sent which divide and end in extremely delicate tracheal capillaries or tracheoles.

Tracheal gills *See* Air tubes in culex.

Tracheoles Same as tracheal capillaries.

Triantominae Suck blood group of lice. *See* Animal Science.

***Tribolium castaneum* Hbst** *See* rust-red flour beetle.

Trichogen cells In cockroach, some of the cells of the epidermis produce movable bristles. Such epidermal cells are known as trichogen cells.

Trichophaga tapetzella This is a tube building clothes moth in which the forewings are black and white. In case of these all the cloth moths, it is the larvae which does all the damage.

Trichoptra It includes insects which as adults have very long jointed antennae and two pairs of large membranous wings clothed with long silky hairs.

Many aquatic larvae build portable cadses, *e.g.*, Caddis flies.

Trinevitermes heimi **Wasm** *See* termites of Sugarcane.

Trochanter *See* legs of cockroach.

Trogoderma granarium **Everts** *See* khapra beetle.

True legs *See* legs of butterflies.

Tsetse fly A small blood sucking fly found in tropical Africa. It transmits parasites into human beings, which cause sleeping sickness and other disorders. The parasites cause a disease in cattle and wild animals called nagana and cattle rearing is severely limited in infested areas. The fly breed in lakes and marshes in W.E. and central Africa, and some of the breeding grounds are sprayed with pesticides.

The Tsetse Fly (*Glossina pulpans*)

Tuber moth of potato (*Solanum tubersum* L.) *Gnorimoschema operculella* Zeel. A brownish grey moth with very minute dark markings on forewings; lays creamy coloured eggs; the larvae make tortuous tunnels in the tuber; damage if done primarily in godown when potatoes are stored and to a slight extent in the field. Keep tubers covered with earth in the field; do not expose potatoes while sowing; godowns should be insect-proof; fumigate infested tubers with Methyl bromide; dust seed potatoes with 5 per cent DDT; cold storage is the most satisfactory method.

Tunga penetrans *See* Animal Science.

***Typhlodromus* sp.** *See* Mites of Mango.

***Tyrophagus assiaticus* Evans** *See* Mites of Mango.

U

Urates in insects Excretory materials in the form of urates are deposited in the fat body. Their presence in the fat body indicates that the fat body performs an important part in excretory system; any excess of waste matter not dealt with by the usual excretory organs, collects in the cells of fat body.

Some of the insects are very beneficial to man. Beetles, short tongued flies, butterflies and some moths are among the valuable insects that pollinate flowers. Thus our many important food plants could not survive without insects. The honeybee produce honey and wax. The silkworm, which is one of the most important insects economically, is worth consideration. Lac insect that lives on cactus belong to the family coccidae, secretes a wax known as shellac. The dye known as cochineal is made from the dried bodies of scale insects, that live on cactus plants.

Useful insects Some of the insects are very useful to man such as Beetles, short tongued flies, butterflies and some moths. The honey bees produce honey and bee wax, the silkworm produces silk, lac-insects (family *Coccideae)* secrete a wax known as shellac. The dye known as coccineal-dye is made from the dried bodies of a scale insect that lives on certain cacti plants.

Utethesia pulchella **L.** *See* dairy caterpillar of Sannhemp.

V

Vasa deferentia The testies in cockroach become mature even when the insect is young. From the testes arise vasadeferentia which are very fine tubes; they continue posterio-ventrally to the hinder part of the abdomen where they open into the upper surface of the ejaculator duct at the base of the mushroom or articulator gland. It consists of numerous tubules which are (*i*) the long peripheral tubules (*ii*) short central tubules (*iii*) short bladder-like outgrowths.

Vector Disease carrying insect such as aphide, jassids and the white flies etc.

Ventilation movement In cockroach, the respiratory is also helped by the ventilation movements, produced by the contraction and expansion of the abdomen.

Ventral diaphragm The previsceral haemocoel is the largest and is bounded dorsally by the pericardium and beneath by the ventral diaphragm. Laterally channels lead into the sternal haemocoel.

Versutu har *See* leaf and fruit eating beetles of apple.

Vertex In insects, the top of the head capsule is known as vertex.

Vertebrate pests Rodents and birds often cause considerable damage to various crops including all grain crops. Rodent species are *Tatera indica, Bendicola bengalansis, Rattus moevigious, Millardia meltada, Mus masculus* and *M. buduga;* are the important rodents that damage our crops. The control measures include trapping, cynogassing of rat burrows and poison baiting with zinc phosphide.

Vesiculae seminales These serve to store sperms in a male cockroach.

Vespoidea *See* Apocrita.

Vine girdler beetle of grape-vine (*Bitis vinifera* L.) *Sthenias girsator* Fabr. Adults make a ring around the stem by girdling it. In severe attack, the plant may die. Cut the attacked branches below the girdling point and burn them.

***Virachola isocrates* Fabr** *See* Anar butterfly of Pomegranate.

Visiting parasites Such as mosquitoes and fleas.

Vitrelle These are the groups of cells lying below the corneagen cells, the inner borer of each of which becomes refractive and the entire refractive structure forms the crystalline cone. Beneath the vitrallae is another group of seven cells, the retinal cells or retinulae, the inner borders of which become refractive each forming a rhabdomere. The group of rabdomeres forms the inner refractive body known as rhabdom.

W

Water fleas *See* Daphnia.

Wax moth It is the worst enemy of bees. Several species of moths cause damage to honeycombs, the two that do the most damage are the Greater Wax Moth (*Galleria mellonella* Linn.) and the Lesser Wax Moth is the worse of the two. The damage is caused by the larvae, which eat through the combs making characteristic tunnels and weave round them a silken web. The bees throw them off when they can get at them, but when the larvae are enveloped in their silken webs they are comparatively safe. The bees apparently dislike the web, for their legs and wings become entangled in it.

Waxing slope A convex slope, which steepens downhill, at the crest of a hill slope. It is the upper slope element in Walter Penck and Lester King's schemes of slope evaluation, lying above the free face in the model slope.

Weevil of mango It lays in the stone of young fruits. It is *Cryptorhynchus (Stenochetus) gravis* and *C. mangiferae.*

Weevil borer of banana (*Musa paradisiaca* L.) *Cosmopolites sordidus* Germ. The weevil lays eggs at the base of stem or on rhizome; grub bores into rhizome tunnelling the root-stock, and travels a little distance up to stem; weevils reproduce all year round and hide in leaf sheath or near sucker on which they feed. Select insect-free rhizomes for planting; spray with 0.16 per cent DDT around the base of the plants/clumps @ 50–75 gallons per acre.

Wheat aphid of wheat (*Triticum sativum* Lam.) *Schizaphis (taxoptera) graminum* (Rs). Sucks juice of young grains and feeds on stem; multiplies very fast. Dust with 3 to 8 per cent BHC at 15 to 20 lb. per acre or spray with 0.05 per cent nicotine at 60–80 gallons per acre.

White ants Same as termites.

White borer of coffee (*Coffea arabica* L.) *Xylotrechus* white borer of *quadripes* Ch. A stout white grub that bores into stem and branches; the beetle which is dark brown with marks on the upper wings lays eggs in cracks on the stem; pupation takes place inside stem. Remove and destroy dead wood and infested parts; remove eggs by s crapping stem; spray with 0.25 per cent DDT @ 2–3 gallons per plant; scrub or swab the stems with 1 per cent BHC suspension.

White flies Pest of Citrus. Many species–*Dialeuroda citir, D. elongata, Aleurolobus citirfolia,* the black spiny fly *Aleurocanthus husaini* and the black fly A. *woglumi.* Control is same as for shoot fly.

White grub *Maladora* sp. and *Anomala demiandata*–commonly attack wheat crop.

Wingcovers *See* tegmina.

Wireworms *Agriotis* spp.–damage many crops.

Woglumi Ashby See Citrus white flies of Citrus Crops.

Woolly aphis *Eriosoma langierum.* It is an insect pest of temperate fruits and is sometimes called American blight. To control, spray of rosin soap on the forms and the roots may be treated with paradichloro benzene put in tunnels few inches deep, covered with soil. A parasite, *Aphelinus mali* is also said to control this pest.

Woolly aphid of apple (*Pyrus malus* L.) *Eriosoma lanigerum* Hauxm. Wooly aphid or American blight lives in colonies on branches, twigs and on main roots below ground, covered with a white cottony wool; a sucking insect that gives birth to young nymphs. Spray aerial forms with fish-oil rosin soap or 0.06 per cent nicotine sulphate or 0.06 per cent malthion or 0.03 per cent parathion @ 2 to 3 gallons per tree; bury paradichloro benzene crystals 1 ft. from the stem of the tree and 4 inches deep in a trench @ 1 to 4 oz. per tree.

Worker cells See honeycomb.

Workers of bees For the formation of combs, they have pocket like wax-secreting glands on the ventral surface of the second to fifth abdominal segment which secrete wax. The wax is passed back between the segments in the form of plats, which are removed from the abdomen by the spines borne on first joints of the tarsus of each hind leg and transferred to the mandibles where it is mixed with saliva and then moulded and fashioned into familiar hexagonal cells. These are sterile females.

X

Xenopsylla cheopis It is the Indian rat flea, is the principal agent in transmitting Bacillus pestis, the cause of bubonic plague.

X. compactuse *See* Shot hole borers of Coffee.

Xyleborus morstatti **Mgdrn** *See* Shot hole borers of Coffee.

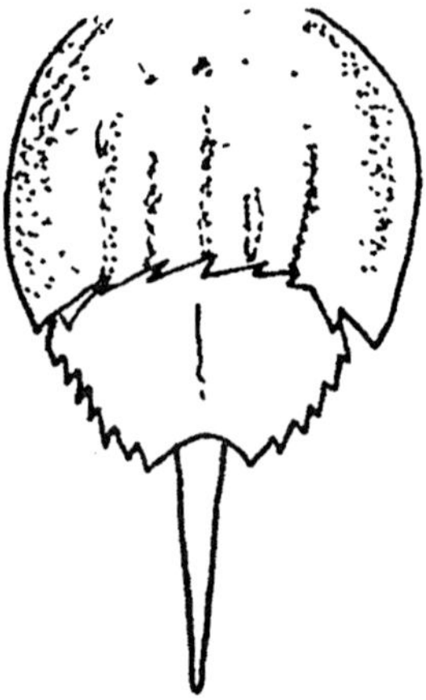

Xiphosura

Y

Yellow meal worm, *Tenebroides mauritanicus* **Linn.** A pale white grub of a flat, shiny-black beetle that attacks hard masses in maida in which the grub lives and feeds; grub stage lasts 10 to 20 months.

Yellow oleander *See Thevetia nerifolia.* A contact insecticide.

Yellow scale *See* scale insects.

Z

Zeuzera coffee **Niet** *See* Red borer of Coffee.

Zeuzera **sp.** See Stem borer of Apple.